Meat Products and Dishes

Sixth supplement to the Fifth Edition of

McCance and Widdowson's

The Composition of Foods

Meat Products and Dishes

Sixth supplement to the Fifth Edition of

McCance and Widdowson's

The Composition of Foods

W. Chan, J. Brown, S. M. Church and D. H. Buss

The Royal Society of Chemistry
and
Ministry of Agriculture, Fisheries and Food

A catalogue record for this book is available from the British Library.

ISBN 0 85404 809 X

Published by the The Royal Society of Chemistry, Cambridge, and the Ministry of Agriculture, Fisheries and Food, London.

Photocomposed by Land and Unwin Ltd, Bugbrooke
Printed in the United Kingdom by the Bath Press, Bath

CONTENTS

		Page
Acknowledgements		vii
Introduction		1
Tables	Symbols and abbreviations	11
	Bacon and ham	12
	Burgers and grillsteaks	24
	Meat pies and pastries	32
	Sausages	40
	Continental style sausages	48
	Other meat products	52
	Meat dishes	68
Appendices	Weight losses on cooking meat products and dishes	107
	Recipes	111
	Ingredient codes of foods used in recipes	137
	Individual fatty acids	140
	Vitamin D fractions	154
	References to tables	156
	Food index	157

ACKNOWLEDGEMENTS

Many people have helped during the preparation of this book.

Most of the new analyses in this book were undertaken by the Laboratory of Government Chemist under the direction of Mr I Lumley and Mrs G Holcombe. Some of the fatty acid analyses were done by Dr M A Jordan at RHM Technology.

We wish to thank numerous manufacturers, retailers and other organisations for information on the range and composition of their products. In particular, we would like to thank Asda Stores Ltd, Burger King UK Ltd, CWS Stores Ltd, Kentucky Fried Chicken GB Ltd, McDonald's Restaurants Ltd, Marks and Spencer plc, Meat and Livestock Commission, Safeway Stores plc, Tesco Stores Ltd and Wimpy International for providing additional data. The Meat and Livestock Commission and the British Chicken Information Service kindly provided the cover photographs.

We would like to express our appreciation for all the help given to us by many people in the Ministry of Agriculture, Fisheries and Food (MAFF) and The Royal Society of Chemistry (RSC) and elsewhere who were involved in the work leading up to the production of this book. In particular, we would like to thank Miss Pratibha Patel (MLC) and Ms Azmina Govindji for their helpful advice; Ms J Alger, Ms S Long and Ms A Sheasby for their work on recipe testings; and Mrs A Philips (RSC) for all her work in the preparation and production of this supplement.

The preparation of this book was overseen by a committee which, besides the authors, comprised Dr A M Fehily (HJ Heinz Company Ltd), Ms J Higgs (Meat and Livestock Commission), Dr J M Hughes (formerly MAFF), Miss A A Paul (MRC Dunn Nutrition Centre), Mrs M de Wet (Northwick Park Hospital), Dr A H Skull (RSC), and Professor D A T Southgate (formerly the Institute of Food Research).

INTRODUCTION

This is the ninth detailed reference book on the nutrients in food, in a series updating and extending the information in McCance and Widdowson's *The Composition of Foods*. It gives the nutrients in commercial meat products including bacon, ham, burgers, pies, sausages, canned, chilled, frozen and long-life products, and meat-based ready meals, together with values for a wide range of meat-based dishes now eaten in and outside the home in Britain. It thus complements the previous supplement which showed the nutrients in carcase meats, poultry, game and offal (Chan *et al.*, 1995).

A large number of new analyses have been undertaken on the increasing range of meat products now available in this country. The values for meat dishes are also new, either because they were not included before or because the information in the fifth edition of *The Composition of Foods* (Holland *et al.*, 1991b) was calculated from the nutritional value of meat given in the fourth edition (Paul and Southgate, 1978) and thus reflected its composition in the early 1970s. There have been substantial changes in the composition of meat since then (see Chan *et al.*, 1995), especially reductions in the amounts of fat both on the carcase itself and after trimming in the shop or in the home. There have also been changes in cooking methods and an increase in the range of poultry dishes. New values have therefore been added or recalculated using the most recent information on the nutrients in all the ingredients. As a result, the number of foods in this supplement has increased to 159 meat products and 127 meat dishes compared with 59 and 24 in the fifth edition. There is also an increase in the number of nutrients for which values are given.

These tables are part of a series produced by The Royal Society of Chemistry (RSC) and the Ministry of Agriculture, Fisheries and Food (MAFF), who have been collaborating since 1987 on the development of a comprehensive and up-to-date database on nutrients in the wide range of foods now available in Britain. The other detailed supplements cover *Cereals and Cereal Products* (Holland *et al.*,1988), *Milk Products and Eggs* (Holland *et al.*,1989), *Vegetables, Herbs and Spices* (Holland *et al.*, 1991a), *Fruit and Nuts* (Holland *et al.*, 1992a), *Vegetable Dishes* (Holland *et al.*, 1992b), *Fish and Fish Products* (Holland *et al.*, 1993), *Miscellaneous Foods* (Chan *et al.*, 1994), and *Meat, Poultry and Game* (Chan *et al.*, 1995). Computerised versions are also available, details of which can be obtained from The Royal Society of Chemistry.

Methods

The selection of foods and the determination of nutrient values follows the general principles used for previous books in this series. All the meat products and meat dishes were chosen to represent as far as possible those most widely available and eaten in Britain at the present time. Most nutrients in most of the meat products were determined by new analyses, although some were derived by interpolation and some were obtained from manufacturers and the scientific literature. A large number of caterers, dietitians, researchers, consumers and recipe books were consulted before the final choice of dishes and recipes was made.

Literature values

The scientific literature as well as manufacturer's information was first reviewed for details of the composition of home-produced and imported meat products. Many of these products are new or have been reformulated, so that even for those foods included in the fifth edition the earlier values could in general not be used, while values from other countries often refer to different products. Manufacturer's information and most literature values were therefore only included where full details of the samples were known; where they were clearly the same as the products now available in British shops or caterers; where suitable methods of analysis had been used; and where the results were available in sufficient detail for a full assessment to be made. The only exceptions were for continental-style sausages, where this supplement includes some values from foreign food tables even though these products may be formulated slightly differently in the UK.

Analyses

Most of the numerous new analyses needed to complete these tables were commissioned by the Ministry of Agriculture, Fisheries and Food from the Laboratory of the Government Chemist (LGC) between 1990 and 1995, with some additional fatty acid analyses carried out by RHM Technology. Up to 10 representative samples of each of the products to be included were bought from a wide variety of supermarkets and other shops, in proportions based upon their share of the market. Items that needed to be cooked were cooked individually by specified methods until done, and any change in weight was recorded. Any cooking instructions on packets were followed, and if two or more methods were given some samples were cooked by each. Blended vegetable oil was used for frying (including stir-frying) unless otherwise specified. When ready meals included rice, this was cooked and weighed separately so that, where appropriate, nutrients could be given for both the meat-based part and the complete meal.

As in previous supplements, the individual samples of each raw or cooked food were combined before analysis. The analytical methods were as described in the fifth edition of *The Composition of Foods* (Holland *et al.*, 1991b). In addition, the individual fatty acids were determined as their methyl esters by capillary gas chromatography and vitamin D and 25-hydroxy vitamin D were determined by quantitative HPLC. For practical reasons, the vitamins in some of the fried and grilled bacon samples, sausages and breaded poultry products were estimated by interpolation from the values in the raw products, usually in proportion to the protein after allowance for cooking losses as in Table 1, except that cholesterol was related to the fat. This reflected the way in which the vitamins and cholesterol were found to vary in the products that were fully analysed. Further details of each estimation can be provided on request.

Recipes

All the values for homemade meat dishes have been derived by calculation from selected recipes, and every attempt was made to ensure that the recipes were as representative as possible. They were obtained from a wide variety of sources including caterers, dietitians, research workers who have conducted quantitative dietary surveys on representative groups of the British population, from standard recipe books, and from consumers. Where more than one recipe is commonly used, for example in different parts of the country, a judgement was made on the most appropriate recipe to use. All the main recipes were then checked and prepared by qualified home economists. Full details of each recipe are in an appendix to this supplement, and users are reminded of the importance of allowing for any differences between these and the recipes they are using or assessing.

The calculations were done as follows. First, each ingredient was prepared and weighed, and the total amount of each nutrient in the uncooked dish was calculated from these weights and the composition of each ingredient taken from the most recent supplements, including this one and the supplement on *Meat, Poultry and Game* (Chan *et al., 1995*). Except where noted, values for meats included any fat normally present on the cut as sold. If the meat was pre-browned or foods were fried, any uptake of fat was recorded and any fat or juices lost were separated and the weights determined. The total weight was then recorded using a scale weighing to about 2g, and the dish was cooked as specified and reweighed to determine any losses including that of water by evaporation during cooking.

As the loss of weight on cooking most recipes was solely from the loss of water, the composition of the cooked dish was calculated as below:

$$\text{Nutrient per 100g of the cooked dish} = \text{Total amount of the nutrient in the raw ingredients} \times \frac{100}{\text{Weight of the cooked dish}} \times (100 - \% \text{ vitamin loss on cooking})^a$$

[a] 100 - % vitamin loss on cooking = % retention

Although there will be little or no loss of fat, protein or minerals during cooking if they are leached into the sauce, juices or liquor and these are eaten as part of the dish, there will still be losses of heat-labile vitamins. The losses that were assumed in the calculations are given in **Table 1**, which is taken from Holland *et al.*, (1991b). In practice, there will be considerable variation about these values

---

Table 1: - Typical percentage losses of vitamins when meat and meat dishes are cooked

	Meat, grilled or fried	Meat dishes[a]
Vitamin A	0	0
Vitamin E	20	20
Thiamin	20	20
Riboflavin	20	20
Niacin	20	20
Vitamin B_6	20	20
Vitamin B_{12}	20	20
Folate	$-$ [b]	50
Pantothenate	20	20
Biotin	10	10
Vitamin C	$-$	50

[a] Some vitamins are lost on heating, but the vitamins (and minerals and fat) that leach into the liquor during cooking will not be lost if the sauce of the gravy is eaten as part of the dish. On average, therefore, the losses in meat dishes are no higher than from grilled or fried meat even though the cooking times are longer.

[b] The amounts of folate in meat are too low to make meaningful calculations of losses.

depending upon the time and temperature of cooking, the nature of the ingredients, and whether conventional or microwave ovens are used, so the values for vitamins in the cooked dishes in this supplement should be treated with some caution.

More information about the principles used in calculating nutrients in recipe dishes, and related examples, are given in the fifth edition of *The Composition of Foods* (Holland *et al.*, 1991b).

Arrangement of the tables

Food groups

For ease of reference, the foods in this book have been listed alphabetically within the following groups: bacon and ham; burgers and grillsteaks; meat pies and pastries; sausages; continental-style sausages; other commercial meat products; and meat dishes. Homemade and commercial versions of similar products have, however, usually been listed together. Where meat products are cooked, values for the raw product are given first.

Numbering system

As in previous supplements, the foods have been numbered in sequence together with a unique two digit prefix. For this supplement, the prefix is 19, so that the full code numbers for raw back bacon rashers and Wiener schnitzel, the first and last foods in this book, are 19-001 and 19-286. These are the numbers that will be used in nutrient databank applications.

Description and number of samples

The information given under this heading includes the source and number of samples taken for analysis, and the source of any literature values. For meat dishes, there is a reference to the recipe given in the appendix at the end of this book.

Nutrients

The nutrient values for each food are shown on four consecutive pages as in most previous books in this series. The presentation follows the established pattern, with the nutrients on three of the pages being similar to those in other supplements, while those on the second page are those most appropriate to (in this supplement) meat products and meat dishes. All values are given per 100 grams of the food as described, but typical weights of selected meat products as bought or served can be found in *Food Portion Sizes* (Ministry of Agriculture, Fisheries and Food, 1993).

Proximates:– The first page for each food shows the amounts of water, total nitrogen, protein, fat, and available carbohydrate expressed (with a few exceptions) as its monosaccharide equivalent. No value is given for edible portion since it is 1.00 for most foods in this supplement. Where it is less than this (mainly for those few samples of bacon bought with rind on), the value is given in a footnote. The protein in each meat product was derived by multiplying the analysed nitrogen values by 6.25 after first subtracting any non-protein nitrogen, and the carbohydrate includes any oligosaccharides from vegetables or other ingredients. Each food's energy value is given both in kilocalories and in kilojoules per 100g, and was derived by multiplying the amounts of protein, fat and carbohydrate by the factors in **Table 2**. The alcohol used in some meat dishes was, however, ignored as it was assumed to have evaporated.

Table 2: - Energy conversion factors used in these tables

	kcal/g	kJ/g
Protein	4	17
Fat	9	37
Available carbohydrate, expressed as monosaccharide	3.75	16

Carbohydrates, fibre and fats: – The second page shows the amounts of starch, total sugars and dietary fibre derived from the other ingredients. Many meat products will contain no carbohydrates or fibre, and for these zero values have been imputed. Where present, carbohydrates have as far as possible been given as their monosaccharide equivalents, while fibre is the actual weight of fibre components (determined by the Englyst and Southgate methods, as in previous supplements). In addition, the total amounts of saturated, monounsaturated (*cis* and *trans* together) and polyunsaturated fatty acids (also *cis* and *trans* together) are given for each meat product and dish, with a further column showing the total *trans* fatty acids separately. The amount of cholesterol is also shown.

Minerals and vitamins: – The range of minerals and vitamins shown is the same as in previous books. Vitamin C is present in many meat products because it is often added as an antioxidant, but it should be noted that from January 1996 isoascorbic acid (D-erythorbic acid), which has little or no vitamin C activity, may be added instead. The amounts of each should therefore be carefully noted in future to avoid overestimation of the nutritional value of the product. Values for carotene and vitamin E have been corrected for the relative activities of the different fractions using the factors given in the fifth edition of *The Composition of Foods* (Holland *et al.*, 1991b), and carotene and retinol values below the limit of detection (usually 5μg per 100g) are given as trace. Vitamin D activity has been taken wherever possible as the amount of cholecalciferol plus five times the amount of any of the more active metabolite 25-hydroxycholecalciferol known or estimated to be present. Where these were measured separately, the amounts are given in an appendix on page 154.

Appendices

There are five appendices in this supplement. The first shows the average loss of weight when various meat products were cooked. The second gives the recipes and measured weight loss for each meat dish that was prepared for this supplement, and the third shows the full nutrient database code numbers for all the ingredients in each dish so that the exact source of the nutrient data used in the recipe calculations can be identified. The final appendices give additional details of the main fatty acids and vitamin D fractions in selected meat products.

Nutrient variability

Almost all foods vary somewhat in nutritional value, and this is particularly important for meat, meat products and meat dishes. For many commercial meat products, the proportions of lean and fat in the meat and the amounts of any other ingredients can vary from one manufacturer to another. There can also be seasonal and other variations according to the cost of the ingredients, as well as gradual changes over time as, for example, manufacturers reduce the amount of fat or salt in their products. Cooking and reheating times and conditions also affect many nutrients. Homemade meat dishes can also vary in composition, because there may be substantial differences in the amounts of meat and other ingredients used, in the proportions of lean and fat in the meat, and in the length and temperature of cooking. All of these influence the composition of the dish as eaten, as can whether a conventional or microwave oven is used. Although care has been taken to ensure that each recipe is as representative as possible of the dish described, note should be taken of any regional or other variation from people systematically using different amounts of ingredients, substituting one ingredient for another, or even omitting one or more ingredients altogether.

Although some reduced fat alternatives have been included, it has not been practicable to give different nutrient values to reflect all these variations. Most values in these tables are therefore average or typical values for each product and dish. They should therefore not be used uncritically: unless the product is similar to that described here, allowance should as far as possible be made for differences in the amounts and nature of the ingredients, including the amount of visible fat on the meat, the cooking methods including the amounts of water and salt used, the length of cooking, and whether or not any fat is removed before or after cooking.

In addition, there are some apparent (but usually small) differences in composition between related products in this supplement which may reflect analytical variations as much as real differences in composition.

The introductions to the fifth edition of *The Composition of Foods* and to the supplement on *Meat, Poultry and Game* give more detailed descriptions of these and other factors that should be taken into account in the proper use of food composition tables. Users of the present supplement are advised to read them and take them to heart.

References to introductory text

Chan, W., Brown, J., and Buss, D.H. (1994) *Miscellaneous Foods*. Fourth supplement to 5th edition of *McCance and Widdowson's The Composition of Foods*. The Royal Society of Chemistry, Cambridge

Chan, W., Brown, J., Lee, S. M., and Buss, D.H. (1995) *Meat, Poultry and Game*. Fifth supplement to 5th edition of *McCance and Widdowson's The Composition of Foods*. The Royal Society of Chemistry, Cambridge

Holland, B., Unwin, I.D., and Buss, D.H. (1988) *Cereals and Cereal Products*. Third supplement to *McCance and Widdowson's The Composition of Foods*, The Royal Society of Chemistry, Cambridge

Holland, B., Unwin, I.D., and Buss, D.H. (1989) *Milk Products and Eggs*. Fourth supplement to *McCance and Widdowson's The Composition of Foods*. The Royal Society of Chemistry, Cambridge

Holland, B., Unwin, I.D., and Buss, D.H. (1991a) *Vegetables, Herbs and Spices*. Fifth supplement to *McCance and Widdowson's The Composition of Foods*, The Royal Society of Chemistry, Cambridge

Holland, B., Welch, A.A., Unwin, I.D., Buss, D.H., Paul, A.A. and Southgate, D.A.T. (1991b) *McCance and Widdowson's The Composition of Foods*, 5th edition, The Royal Society of Chemistry, Cambridge

Holland, B., Unwin, I.D., and Buss, D.H. (1992a) *Fruit and Nuts*. First supplement to 5th edition of *McCance and Widdowson's The Composition of Foods*. The Royal Society of Chemistry, Cambridge

Holland, B., Welch, A.A., and Buss, D.H. (1992b) *Vegetable Dishes*. Second supplement to 5th edition of *McCance and Widdowson's The Composition of Foods*. The Royal Society of Chemistry, Cambridge

Holland, B., Brown, J., and Buss, D.H. (1993) *Fish and Fish products*. Third supplement to 5th edition of *McCance and Widdowson's The Composition of Foods*. The Royal Society of Chemistry, Cambridge

Ministry of Agriculture, Fisheries and Food (1993) *Food Portion Sizes*, 2nd edition, HMSO, London

Paul, A.A. and Southgate, D.A.T. (1978) *McCance and Widdowson's The Composition of Foods*, 4th edition, HMSO, London

The
Tables

Symbols and abbreviations used in the tables

Symbols

0	None of the nutrient is present
Tr	Trace
N	The nutrient is present in significant quantities but there is no reliable information on the amount
( )	Estimated value
Italic text	Carbohydrate and starch estimated 'by difference', and energy values based upon these quantities

Abbreviations

Satd	Saturated
Monounsatd	Monounsaturated
Polyunsatd	Polyunsaturated
Trypt	Tryptophan
equiv	equivalents

Bacon and ham

Composition of food per 100g

No. 19-	Food	Description and main data sources	Water g	Total Nitrogen g	Protein g	Fat g	Carbo-hydrate g	Energy value kcal	kJ
1	**Bacon rashers, back**, *raw*	10 samples; smoked and unsmoked, loose and prepacked British, Danish and Dutch bacon[a]	63.9	2.64	16.5	16.5	0	215	891
2	-, *dry-fried*	10 samples; smoked and unsmoked, loose and prepacked British, Danish and Dutch bacon	49.7	3.87	24.2	22.0	0	295	1225
3	-, *grilled*	15 samples; smoked and unsmoked, loose and prepacked British, Danish and Dutch bacon	50.4	3.71	23.2	21.6	0	287	1194
4	-, *grilled crispy*	10 samples; smoked and unsmoked, loose and prepacked British, Danish and Dutch bacon	37.8	5.76	36.0	18.8	0	313	1308
5	-, *microwaved*	15 samples; smoked and unsmoked, loose and prepacked British, Danish and Dutch bacon	45.5	3.87	24.2	23.3	0	307	1274
6	-, *dry-cured, grilled*	7 samples; smoked and unsmoked, loose and prepacked British and Irish bacon	50.7	4.54	28.4	15.9	0	257	1071
7	-, *fat trimmed, raw*	24 samples, back fat removed. MLC data and calculation from No 8	69.5	3.01	18.8	6.7	0	136	568
8	-, -, *grilled*	15 samples; smoked and unsmoked, loose and prepacked British, Danish and Dutch bacon, back fat removed	56.2	4.11	25.7	12.3	0	214	892
9	-, *reduced salt, grilled*	6 samples; smoked and unsmoked, loose and prepacked British and Danish bacon	51.6	3.86	24.1	20.6	0	282	1172
10	-, *smoked, grilled*	9 samples; loose and prepacked British and Danish bacon	50.5	3.74	23.4	22.1	0	293	1216
11	-, *sweetcure, grilled*	9 samples; smoked and unsmoked, loose and prepacked British and Danish bacon	52.8	3.81	23.8	17.4	1.6	258	1074

[a] If bought with rind, edible proportion is 0.97

Bacon and ham

Composition of food per 100g

No. 19-	Food	Starch g	Total sugars g	Dietary fibre		Fatty acids				Cholesterol mg
				Southgate method g	Englyst method g	cis & trans			Total trans g	
						Satd g	Mono-unsatd g	Poly-unsatd g		
1	**Bacon rashers, back**, *raw*	0	0	0	0	6.2	6.9	2.2	0.1	53
2	-, *dry-fried*	0	0	0	0	8.3	9.2	2.8	0.1	65
3	-, *grilled*	0	0	0	0	8.1	9.0	2.8	0.1	75
4	-, *grilled crispy*	0	0	0	0	7.1	7.9	2.4	0.1	68
5	-, *microwaved*	0	0	0	0	8.8	9.8	3.0	0.1	84
6	-, *dry-cured, grilled*	0	0	0	0	6.0	6.7	2.1	Tr	57
7	-, *fat trimmed, raw*	0	0	0	0	2.5	2.8	0.9	Tr	(31)
8	-, -, *grilled*	0	0	0	0	4.6	5.2	1.6	0.1	44
9	-, *reduced salt, grilled*	0	0	0	0	7.8	8.7	2.7	0.1	74
10	-, *smoked, grilled*	0	0	0	0	8.3	9.3	2.9	0.1	80
11	-, *sweetcure, grilled*	0	1.6	0	0	6.6	7.3	2.2	0.1	63

Inorganic constituents per 100g food

No. 19-	Food	Na	K	Ca	Mg	P	Fe	Cu	Zn	Cl	Mn	Se	I
							mg					µg	
1	**Bacon rashers, back**, *raw*	1540	300	5	17	150	0.4	0.06	1.2	2350	0.01	8	5
2	-, *dry-fried*	1910	360	6	21	180	0.6	0.06	1.9	(3510)	0.01	18	7
3	-, *grilled*	1880	340	7	21	180	0.6	0.05	1.7	2780	0.01	12	7
4	-, *grilled crispy*	(2700)	510	10	32	300	1.1	0.10	3.1	(3510)	0.01	18	11
5	-, *microwaved*	2330	360	8	23	200	0.7	0.06	2.0	(2360)	0.01	12	7
6	-, *dry-cured, grilled*	(2140)	400	9	26	240	0.8	0.08	2.4	(2770)	0.01	14	9
7	-, *fat trimmed, raw*	(1350)	(250)	(6)	(16)	(150)	(0.5)	(0.05)	(1.5)	(1740)	(0.01)	(9)	(6)
8	-, -, *grilled*	(1930)	360	8	23	210	0.7	0.07	2.2	(2500)	0.01	13	8
9	-, *reduced salt, grilled*	1130	340	7	22	200	0.7	0.09	2.1	1500	0.01	12	7
10	-, *smoked, grilled*	(1760)	330	7	21	190	0.7	0.06	2.2	(2280)	0.01	12	7
11	-, *sweetcure, grilled*	(1790)	340	7	21	200	0.8	0.09	2.4	(2320)	0.01	12	7

No. 19-	Food	Retinol µg	Carotene µg	Vitamin D µg	Vitamin E mg	Thiamin mg	Ribo-flavin mg	Niacin mg	Trypt 60 mg	Vitamin B6 mg	Vitamin B12 µg	Folate µg	Panto-thenate mg	Biotin µg	Vitamin C mg
1	**Bacon rashers, back**, *raw*	Tr	Tr	0.3	0.02	0.63	0.11	5.6	2.6	0.46	Tr	3	1.00	2	1
2	-, *dry-fried*	Tr	Tr	0.6	0.07	0.86	0.14	6.8	4.4	0.53	1	2	1.26	5	Tr
3	-, *grilled*	Tr	Tr	0.6	0.07	1.16	0.15	7.2	3.8	0.52	1	5	1.24	3	Tr
4	-, *grilled crispy*	Tr	Tr	1.0	0.10	1.38	0.24	10.8	6.6	0.71	1	4	1.34	5	Tr
5	-, *microwaved*	Tr	Tr	0.6	0.07	1.10	0.16	7.9	4.4	0.55	1	2	1.26	5	Tr
6	-, *dry-cured, grilled*	Tr	Tr	0.8	0.08	1.09	0.19	8.5	3.5	0.56	1	3	1.48	6	Tr
7	-, *fat trimmed, raw*	Tr	Tr	(0.5)	(0.05)	(0.68)	(0.12)	(5.4)	(3.3)	(0.35)	(1)	(2)	(0.93)	(4)	Tr
8	-, -, *grilled*	Tr	Tr	0.7	0.07	0.98	0.17	7.7	4.7	0.50	1	3	1.34	5	Tr
9	-, *reduced salt, grilled*	Tr	Tr	0.6	0.07	0.92	0.16	7.2	4.4	0.47	1	2	1.25	5	Tr
10	-, *smoked, grilled*	Tr	Tr	0.6	0.07	0.90	0.16	7.0	4.3	0.46	1	2	1.22	5	Tr
11	-, *sweetcure, grilled*	Tr	Tr	0.6	0.07	0.91	0.16	7.1	4.4	0.47	1	3	1.24	5	Tr

Bacon and ham *continued*

Composition of food per 100g

No. 19-	Food	Description and main data sources	Water g	Total Nitrogen g	Protein g	Fat g	Carbo-hydrate g	Energy value kcal	kJ
12	**Bacon rashers, back,** 'tendersweet', *grilled*	10 samples; smoked and unsmoked, loose and prepacked British bacon	55.4	4.22	26.4	11.9	Tr	213	889
13	-, **middle**, *raw*	9 samples; smoked and unsmoked, loose and prepacked British and Danish bacon[a]	59.9	2.43	15.2	20.0	0	241	998
14	-, -, *fried*	9 samples; smoked and unsmoked, loose and prepacked British and Danish bacon	43.6	3.74	23.4	28.5	0	350	1452
15	-, -, *grilled*	9 samples; smoked and unsmoked, loose and prepacked British and Danish bacon	47.8	3.97	24.8	23.1	0	307	1276
16	-, **streaky**, *raw*	10 samples; smoked and unsmoked, loose and prepacked British and Danish bacon[b]	57.3	2.53	15.8	23.6	0	276	1142
17	-, -, *fried*	10 samples; smoked and unsmoked, loose and prepacked British and Danish bacon	45.1	3.81	23.8	26.6	0	335	1389
18	-, -, *grilled*	10 samples; smoked and unsmoked, loose and prepacked British and Danish bacon	44.0	3.81	23.8	26.9	0	337	1400
19	**Bacon loin steaks**, *grilled*	7 samples; smoked and unsmoked, loose and prepacked Danish bacon	60.6	4.14	25.9	9.7	0	191	799
20	**Ham, gammon joint**, *raw*	10 samples; smoked and unsmoked, prepacked British and Danish gammon[a]	68.6	2.80	17.5	7.5	0	138	575
21	-, *boiled*	10 samples; smoked and unsmoked, prepacked British and Danish gammon	61.2	3.73	23.3	12.3	0	204	851
22	-, **gammon rashers**, *grilled*	5 samples; unsmoked British gammon	58.2	4.40	27.5	9.9	0	199	834

[a] If bought with rind, edible proportion is 0.92   [b] If bought with rind, edible proportion is 0.91

No. Food	Starch	Total sugars	Dietary fibre Southgate method	Dietary fibre Englyst method	Fatty acids Satd	Fatty acids cis & trans Mono-unsatd	Fatty acids cis & trans Poly-unsatd	Total trans	Cholesterol
19-	g	g	g	g	g	g	g	g	mg
12 **Bacon rashers, back,** 'tendersweet', *grilled*	0	Tr	0	0	4.5	5.0	1.5	Tr	43
13 -, **middle**, *raw*	0	0	0	0	7.3	8.6	2.5	0.1	58
14 -, -, *fried*	0	0	0	0	9.8	11.9	4.8	0.1	84
15 -, -, *grilled*	0	0	0	0	8.4	10.0	3.0	0.1	83
16 -, **streaky**, *raw*	0	0	0	0	8.2	10.2	3.5	0.1	65
17 -, -, *fried*	0	0	0	0	9.1	11.1	4.5	2.6	78
18 -, -, *grilled*	0	0	0	0	9.8	11.5	3.7	0.1	90
19 **Bacon loin steaks**, *grilled*	0	0	0	0	3.5	4.1	1.4	Tr	69
20 **Ham, gammon joint**, *raw*	0	0	0	0	2.5	3.3	1.2	Tr	23
21 -, *boiled*	0	0	0	0	4.1	5.4	1.9	Tr	83
22 -, **gammon rashers**, *grilled*	0	0	0	0	3.4	4.1	1.7	0.1	83

Bacon and ham *continued*

Inorganic constituents per 100g food

No. 19-	Food	mg										μg	
		Na	K	Ca	Mg	P	Fe	Cu	Zn	Cl	Mn	Se	I
12	**Bacon rashers, back,** 'tendersweet', *grilled*	(1990)	370	8	24	220	0.8	0.09	2.4	(2320)	0.01	12	7
13	-, **middle**, *raw*	1480	260	6	16	160	0.4	0.08	1.5	1560	0.01	7	6
14	-, -, *fried*	1840	350	10	20	210	0.7	0.07	2.3	(2280)	0.01	12	7
15	-, -, *grilled*	1960	350	8	21	220	0.7	0.07	2.2	2050	0.01	11	8
16	-, **streaky**, *raw*	1260	250	6	15	140	0.5	0.06	1.5	1500	0.01	7	7
17	-, -, *fried*	(1880)	350	7	21	200	0.7	0.07	2.1	(2320)	0.01	10	7
18	-, -, *grilled*	1680	330	9	20	180	0.8	0.15	2.5	2630	0.01	11	6
19	**Bacon loin steaks**, *grilled*	1480	370	6	23	220	0.6	0.07	1.9	(2520)	0.01	13	8
20	**Ham, gammon joint**, *raw*	(880)	190	7	17	130	0.6	0.08	1.5	(1980)	0.01	11	7
21	-, *boiled*	1180	250	9	18	170	0.8	0.10	2.1	(2640)	0.01	12	9
22	-, **gammon rashers**, *grilled*	1930	380	8	26	230	0.8	0.09	2.2	(2680)	0.02	14	8

Bacon and ham continued

No. Food 19-	Retinol µg	Carotene µg	Vitamin D µg	Vitamin E mg	Thiamin mg	Ribo-flavin mg	Niacin mg	Trypt 60 mg	Vitamin B6 mg	Vitamin B12 µg	Folate µg	Panto-thenate mg	Biotin µg	Vitamin C mg
12 **Bacon rashers, back,** 'tendersweet', *grilled*	Tr	Tr	0.6	0.08	1.01	0.18	7.9	4.8	0.51	1	3	1.37	5	Tr
13 -, **middle,** *raw*	Tr	Tr	0.5	0.07	0.58	0.11	4.9	2.5	0.41	1	5	0.99	3	Tr
14 -, -, *fried*	Tr	Tr	0.6	N	0.64	0.14	7.3	4.3	0.46	1	6	1.22	5	Tr
15 -, -, *grilled*	Tr	Tr	0.6	0.13	0.77	0.17	7.5	5.1	0.42	1	3	1.27	6	Tr
16 -, **streaky,** *raw*	Tr	Tr	0.9	0.07	0.45	0.14	4.7	2.7	0.30	1	3	0.92	2	Tr
17 -, -, *fried*	Tr	Tr	0.6	N	0.75	0.14	7.1	4.4	0.47	1	1	1.24	5	Tr
18 -, -, *grilled*	Tr	Tr	0.7	0.07	0.70	0.17	6.3	4.3	0.40	1	3	1.22	4	Tr
19 **Bacon loin steaks,** *grilled*	Tr	Tr	0.7	0.08	1.19	0.13	7.8	4.8	0.58	1	3	1.35	5	Tr
20 **Ham, gammon joint,** *raw*	Tr	Tr	0.6	0.06	0.44	0.13	5.3	2.9	0.43	Tr	4	1.07	2	Tr
21 -, *boiled*	Tr	Tr	0.8	0.08	0.58	0.16	5.4	3.9	0.42	Tr	3	1.43	2	Tr
22 -, **gammon rashers,** *grilled*	Tr	Tr	0.8	0.08	1.16	0.18	6.4	5.5	0.16	1	3	1.43	6	Tr

No. Food 19-	Description and main data sources	Water g	Total Nitrogen g	Protein g	Fat g	Carbo-hydrate g	Energy value kcal	Energy value kJ
23 **Ham**	10 samples, 9 brands; loose and prepacked including honey roast and smoked ham. Added water 10-15%	73.2	2.94	18.4	3.3	1.0	107	451
24 -, canned	10 samples, 8 brands	74.0	2.64	16.5	4.5	0.1	107	449
25 -, Parma	Suppliers' data	N	4.35	27.2	12.7	Tr	223	932
26 -, premium	10 samples, 7 brands; loose and prepacked including honey glazed and smoked ham. No added water	69.7	3.39	21.2	5.0	0.5	132	553
27 **Pork shoulder**, cured, slices	8 samples, 7 brands	75.2	2.70	16.9	3.6	0.9	103	435

Bacon and ham *continued*

No. 19-	Food	Starch g	Total sugars g	Dietary fibre Southgate method g	Dietary fibre Englyst method g	Fatty acids Satd g	Fatty acids cis & trans Mono-unsatd g	Fatty acids cis & trans Poly-unsatd g	Total trans g	Cholesterol mg
23	**Ham**	0	1.0	0	0	1.1	1.5	0.5	Tr	58
24	-, canned	0	0.1	0	0	1.6	2.0	0.4	Tr	33
25	-, Parma	0	Tr	Tr	Tr	4.3	N	N	N	N
26	-, premium	0	0.5	0	0	1.7	2.2	0.8	Tr	58
27	**Pork shoulder**, cured, slices	0.1	0.8	0	0	1.2	1.6	0.6	Tr	58

Bacon and ham *continued*

Inorganic constituents per 100g food

No. 19-	Food	Na	K	Ca	Mg	P	Fe	Cu	Zn	Cl	Mn	Se	I
						mg						µg	
23	**Ham**	1200	340	7	24	340	0.7	0.12	1.8	1470	0.01	11	5
24	-, canned	1470	200	32	18	270	1.2	0.10	2.0	1770	0.02	(8)	11
25	-, Parma	2000	N	N	N	N	N	N	N	N	N	N	N
26	-, premium	1050	320	6	22	240	0.8	0.07	2.1	1420	0.01	12	5
27	**Pork shoulder**, cured, slices	1000	320	6	19	300	1.0	0.08	2.5	1280	0.01	10	5

Bacon and ham *continued*

Vitamins per 100g food

No. Food	Retinol	Carotene	Vitamin D	Vitamin E	Thiamin	Ribo-flavin	Niacin	Trypt/60	Vitamin B6	Vitamin B12	Folate	Panto-thenate	Biotin	Vitamin C
19-	µg	µg	µg	mg	mg	mg	mg	mg	mg	µg	µg	mg	µg	mg
23 **Ham**	Tr	Tr	N	0.04	0.80	0.17	6.5	3.1	0.61	1	19	1.03	3	Tr
24 -, canned	Tr	Tr	N	0.08	0.29	0.24	2.5	3.0	0.21	Tr	2	0.60	1	19
25 -, Parma	N	N	N	N	N	N	N	N	N	N	N	N	N	N
26 -, premium	Tr	Tr	N	(0.05)	0.27	0.20	4.5	N	0.45	Tr	6	N	N	Tr
27 **Pork shoulder**, cured, slices	Tr	Tr	N	N	N	N	N	N	N	N	N	N	N	Tr

Burgers and grillsteaks

Composition of food per 100g

No. 19-	Food	Description and main data sources	Water g	Total Nitrogen g	Protein g	Fat g	Carbo-hydrate g	Energy value kcal	kJ
28	**Beefburgers**, chilled/frozen, *raw*	8 samples, 3 brands. 98-99% meat	56.1	2.74	17.1	24.7	0.1	291	1206
29	-, *fried*	8 samples, 3 brands	46.2	4.56	28.5	23.9	0.1	329	1370
30	-, *grilled*	8 samples, 3 brands	47.9	4.24	26.5	24.4	0.1	326	1355
31	homemade, *fried*	Recipe	49.0	3.81	23.8	22.7	1.0[a]	303	1261
32	-, -, with bun	Recipe	45.9	3.36	21.0	19.3	10.4[b]	297	1238
33	homemade, *grilled*	Recipe	50.0	3.98	24.9	20.4	1.0[a]	287	1194
34	-, -, with bun	Recipe	46.7	3.50	21.9	17.4	10.4[b]	283	1183
35	low fat, chilled/frozen, *raw*	11 samples, 4 brands. 80% meat	65.8	2.86	17.9	9.5	0.2	158	659
36	-, -, *fried*	11 samples, 4 brands	58.0	3.78	23.6	10.8	0.4	193	807
37	-, -, *grilled*	11 samples, 4 brands	60.0	3.63	22.7	9.5	0.5	178	745
38	**Beefburgers in gravy**, canned	10 samples, 4 brands	69.6	1.94	12.1	11.5	5.1	171	713
39	**Big Mac**	Manufacturer's data (McDonald's). Portion includes two beefburgers, bun, sauce, cheese, lettuce, onions and pickles	N	2.03	12.7	12.7	18.0	238	996
40	**Cheeseburger**, takeaway	Manufacturers' data and calculation from ingredient proportions. Includes beefburger, bun, cheese, mustard, ketchup, onions and pickles	N	2.38	14.9	12.4	23.9	267	1118
41	**Chicken burger**, takeaway	Manufacturers' data. Portion includes chicken burger, bun, lettuce, and mayonnaise	N	2.00	12.5	10.8	23.4	267	1118

[a] Includes 0.3g oligosaccharides per 100g food

[b] Includes 0.2g oligosaccharides per 100g food

No. 19-	Food	Starch g	Total sugars g	Dietary fibre Southgate method g	Dietary fibre Englyst method g	Satd g	Fatty acids cis & trans Mono-unsatd g	Fatty acids cis & trans Poly-unsatd g	Total trans g	Cholesterol mg
28	**Beefburgers**, chilled/frozen, *raw*	Tr	0.1	0	0	11.1	11.3	0.7	1.4	76
29	-, *fried*	Tr	0.1	0	0	10.7	10.8	0.8	0.8	96
30	-, *grilled*	Tr	0.1	0	0	10.9	11.2	0.7	1.4	(75)
31	homemade, *fried*	Tr	0.8	0.2	0.2	8.7	9.7	2.2	0.9	110
32	-, -, with bun	9.0	1.1	0.9	(0.4)	7.3	8.0	2.0	0.7	90
33	homemade, *grilled*	Tr	0.8	0.2	0.2	8.8	8.9	0.7	0.9	115
34	-, -, with bun	9.0	1.1	0.9	(0.4)	7.3	7.5	0.8	0.8	94
35	low fat, chilled/frozen, *raw*	Tr	0.2	0	0	4.4	4.2	0.3	0.2	64
36	-, -, *fried*	Tr	0.4	0	0	5.0	4.7	0.4	0.3	(70)
37	-, -, *grilled*	Tr	0.5	0	0	4.4	4.2	0.3	0.2	(64)
38	**Beefburgers in gravy**, canned	4.2	0.9	Tr	Tr	4.8	5.2	0.6	0.4	40
39	**Big Mac**	14.6	3.4	N	N	6.2	N	N	N	N
40	**Cheeseburger**, takeaway	18.3	5.6	(1.6)	(0.7)	(6.0)	(5.4)	(0.9)	(0.3)	(44)
41	**Chicken burger**, takeaway	N	N	Tr	Tr	N	N	N	N	N

Burgers and grillsteaks

Inorganic constituents per 100g food

No. 19-	Food	Na	K	Ca	Mg	P	Fe	Cu	Zn	Cl	Mn	Se	I
							mg					µg	
28	**Beefburgers**, chilled/frozen, *raw*	290	290	7	16	150	1.7	0.12	3.8	350	0.02	8	(8)
29	-, *fried*	470	420	12	26	240	2.8	0.13	6.3	570	0.02	(10)	(13)
30	-, *grilled*	400	380	10	22	210	2.5	0.13	6.1	520	0.02	(9)	(12)
31	homemade, *fried*	330	320	20	22	210	1.9	0.02	4.6	450	0.06	9	17
32	-, -, with bun	370	280	41	23	200	1.9	0.04	3.8	530	0.14	(13)	17
33	homemade, *grilled*	340	340	21	23	220	1.9	0.02	4.8	470	0.07	1	18
34	-, -, with bun	380	300	42	24	200	2.0	0.04	4.0	550	0.14	(13)	18
35	low fat, chilled/frozen, *raw*	630	330	58	20	170	2.7	0.12	4.5	970	0.10	N	N
36	-, -, *fried*	(830)	(440)	(80)	(30)	(230)	(3.6)	(0.16)	(5.9)	(1280)	(0.13)	N	N
37	-, -, *grilled*	(800)	(420)	(75)	(25)	(220)	(3.4)	(0.15)	(5.7)	(1230)	(0.13)	N	N
38	**Beefburgers in gravy**, canned	680	130	50	14	230	1.8	0.11	1.3	940	0.11	N	N
39	**Big Mac**	550	190	55	N	N	0.9	N	N	N	N	N	N
40	**Cheeseburger**, takeaway	750	210	85	(27)	(230)	1.1	(0.13)	(3.0)	(920)	(0.23)	(15)	(15)
41	**Chicken burger**, takeaway	560	190	19	N	N	0.4	N	N	N	Tr	N	N

Burgers and grillsteaks

No. 19-	Food	Retinol µg	Carotene µg	Vitamin D µg	Vitamin E mg	Thiamin mg	Ribo-flavin mg	Niacin mg	Trypt 60 mg	Vitamin B6 mg	Vitamin B12 µg	Folate µg	Panto-thenate mg	Biotin µg	Vitamin C mg
28	**Beefburgers**, chilled/frozen, *raw*	Tr	Tr	1.2	0.28	0.01	0.15	3.5	2.5	0.28	2	9	0.78	1	0
29	-, *fried*	Tr	Tr	(1.9)	0.54	Tr	0.22	5.5	4.3	0.31	3	8	0.85	2	0
30	-, *grilled*	Tr	Tr	(1.8)	0.39	0.01	0.20	5.1	4.0	0.31	3	10	0.84	2	0
31	homemade, *fried*	21	Tr	0.7	(0.29)	0.08	0.16	5.3	4.5	0.37	2	12	0.61	3	0
32	-, -, with bun	17	Tr	0.6	0.23	0.11	0.15	4.6	4.0	0.31	2	19	0.55	3	0
33	homemade, *grilled*	22	Tr	0.7	(0.30)	0.08	0.17	5.6	4.7	0.38	2	12	0.64	3	0
34	-, -, with bun	18	Tr	0.6	(0.24)	0.11	0.15	4.8	4.2	0.32	2	19	0.58	3	0
35	low fat, chilled/frozen, *raw*	Tr	Tr	N	0.08	0.03	0.17	3.9	3.8	0.26	3	6	0.66	2	0
36	-, -, *fried*	Tr	Tr	N	N	(0.04)	(0.22)	(5.1)	(2.9)	(0.34)	(4)	(8)	(0.87)	(3)	0
37	-, -, *grilled*	N	Tr	N	(0.10)	(0.04)	(0.22)	(4.9)	(4.9)	(0.33)	(4)	(8)	(0.84)	(3)	0
38	**Beefburgers in gravy**, canned	N	Tr	N	N	0.02	0.10	1.5	N	0.18	1	5	N	N	0
39	**Big Mac**	N	N	N	N	0.14	0.12	2.2	N	0.17	1	N	N	N	N
40	**Cheeseburger**, takeaway	(24)	(23)	(0.3)	(0.26)	0.17	0.18	2.2	(2.8)	0.19	2	(23)	(0.46)	(1)	N
41	**Chicken burger**, takeaway	N	N	N	N	0.26	0.07	4.3	N	0.28	N	N	N	N	N

Burgers and grillsteaks continued

Composition of food per 100g

No. 19-	Food	Description and main data sources	Water g	Total Nitrogen g	Protein g	Fat g	Carbo-hydrate g	Energy value kcal	kJ
42	**Economy burgers,** frozen, *raw*	10 samples, 6 brands containing onion. 60% meat	57.1	2.19	13.7	21.2	4.0	261	1083
43	-, *grilled*	10 samples, 6 brands containing onion	50.8	2.53	15.8	19.3	9.7[a]	273	1138
44	**Grillsteaks, beef,** chilled/frozen, *raw*	10 samples, 7 brands. 95-98% meat	57.2	2.78	17.4[b]	23.9[b]	0.2[b]	285[b]	1183[b]
45	-, -, *fried*	10 samples, 7 brands	48.5	4.02	25.1	22.2	0.5	302	1256
46	-, -, *grilled*	10 samples, 7 brands	50.1	3.54	22.1	23.9	0.5	305	1268
47	**Hamburger,** takeaway	Manufacturers' data and calculation from ingredient proportions. Portion includes bun, beefburger, mustard, ketchup, onions and pickles	N	2.22	13.9	9.6	26.9	243	1022
48	**Quarterpounder,** takeaway	Manufacturer's data (McDonald's). Portion includes a quarter pound beefburger, bun, ketchup, mustard, onions and pickles	N	2.34	14.6	13.6	18.3	249	1044
49	**Steaklets,** frozen, *raw*	10 samples	55.8	2.45	15.3	24.3	3.4	293	1214
50	**Whopper burger**	Manufacturer's data (Burger King) and calculation from ingredient proportions. Portion includes bun, beefburger, mayonnaise, lettuce, tomato, ketchup, onions and pickles	(52.5)	1.70	10.6	14.1	18.5	239	998

[a] Includes 0.1g oligosaccharides per 100g food   [b] Lamb grills contain 15.4g protein, 25.0g fat, 1.9g carbohydrate, 294 kcal, 1217 kJ per 100g food

Burgers and grillsteaks *continued*

No. 19-	Food	Starch g	Total sugars g	Dietary fibre Southgate method g	Dietary fibre Englyst method g	Satd g	Fatty acids cis & trans Mono-unsatd g	Fatty acids cis & trans Poly-unsatd g	Total trans g	Cholesterol mg
42	**Economy burgers**, frozen, *raw*	3.1	0.9	N	0.9	8.0	9.4	2.1	0.8	(92)
43	-, *grilled*	8.9	0.7	N	0.8	7.3	8.6	1.9	0.7	84
44	**Grillsteaks, beef,** chilled/frozen, *raw*	0.2	Tr	N	(0.6)	10.8	10.7	0.8	0.7	75
45	-, -, *fried*	Tr	0.5	Tr	Tr	10.0	10.0	0.8	(0.7)	88
46	-, -, *grilled*	Tr	0.5	Tr	Tr	10.8	10.7	0.8	(0.7)	88
47	**Hamburger**, takeaway	21.2	5.7	(1.8)	(0.8)	(4.0)	(4.2)	(0.8)	(0.2)	(40)
48	**Quarterpounder**, takeaway	13.4	4.9	N	N	6.4	N	N	N	N
49	**Steaklets**, frozen, *raw*	3.1	0.3	N	0.6	10.1	11.6	0.7	0.9	58
50	**Whopper burger**	15.4	3.1	(1.6)	(0.8)	(4.5)	(5.1)	(4.0)	(0.3)	(36)

Burgers and grillsteaks *continued*

Inorganic constituents per 100g food

No. 19-	Food	Na	K	Ca	Mg	P	Fe (mg)	Cu	Zn	Cl	Mn	Se (µg)	I
42	**Economy burgers**, frozen, *raw*	590	210	32	18	170	2.1	0.11	2.5	830	0.13	N	N
43	-, *grilled*	800	270	110	25	200	2.5	0.15	1.2	(1200)	0.25	N	N
44	**Grillsteaks, beef,** chilled/frozen, *raw*	570[a]	280	18	18	180	2.2	0.12	4.3	710	0.02	N	N
45	-, *fried*	800	400	19	22	220	2.7	0.11	5.2	1100	0.02	N	N
46	-, *grilled*	710	360	18	19	190	2.4	0.10	4.7	980	0.02	(3)	N
47	**Hamburger**, takeaway	620	210	40	(28)	(170)	1.2	(0.12)	(3.0)	(900)	(0.25)	(16)	(13)
48	**Quarterpounder**, takeaway	450	250	28	N	N	1.2	N	N	N	N	N	N
49	**Steaklets**, frozen, *raw*	400	190	26	14	170	1.7	0.04	3.0	420	0.06	N	N
50	**Whopper burger**	270	(230)	(50)	(20)	(130)	(1.8)	(0.09)	(2.2)	(670)	(0.21)	(12)	(13)

[a] Lamb grills contain 500mg Na per 100g food

30

Burgers and grillsteaks *continued*

No. 19-	Food	Retinol µg	Carotene µg	Vitamin D µg	Vitamin E mg	Thiamin mg	Ribo-flavin mg	Niacin mg	Trypt 60 mg	Vitamin B_6 mg	Vitamin B_{12} µg	Folate µg	Panto-thenate mg	Biotin µg	Vitamin C mg
42	**Economy burgers**, frozen, *raw*	(4)	Tr	N	N	0.07	0.23	2.6	N	0.17	2	12	N	N	Tr
43	-, *grilled*	5	Tr	N	N	0.07	0.07	3.8	N	0.17	2	24	N	N	Tr
44	**Grillsteaks, beef**, chilled/frozen, *raw*	Tr	Tr	N	0.13	0.14	0.15	3.6	2.3	0.19	3	4	0.67	2	Tr
45	-, -, *fried*	Tr	Tr	N	N	(0.15)	(0.16)	(4.7)	(5.3)	(0.20)	(3)	N	(0.72)	(2)	Tr
46	-, -, *grilled*	Tr	Tr	N	0.15	0.13	0.14	4.2	4.7	0.18	3	N	0.63	2	Tr
47	**Hamburger**, takeaway	N	Tr	(0.3)	(0.22)	0.19	0.12	2.5	(2.7)	(0.18)	(1)	(24)	(0.48)	(1)	N
48	**Quarterpounder**, takeaway	N	N	N	N	0.16	0.13	3.1	N	0.24	3	N	N	N	N
49	**Steaklets**, frozen, *raw*	Tr	Tr	N	N	0.01	0.12	2.0	N	0.22	2	3	N	N	0
50	**Whopper burger**	(7)	(94)	(0.2)	(1.80)	(0.15)	(0.11)	(2.4)	(2.0)	(0.13)	(1)	(25)	(0.40)	(1)	(2)

Meat pies and pastries

Composition of food per 100g

No. 19-	Food	Description and main data sources	Water g	Total Nitrogen g	Protein g	Fat g	Carbohydrate g	Energy value kcal	kJ
51	**Beef pie**, chilled/frozen, *baked*	20 samples, 6 brands	48.3	1.17	7.3	20.8	22.1	299	1247
52	**Beef steak pudding**, homemade	Recipe	58.0	1.86	11.6	10.5	18.8[a]	211	887
53	**Bridie/Scotch pie**, individual	10 samples	56.7	1.18	7.4	10.5	20.9	202	849
54	**Chicken and mushroom pie**, single crust, homemade	Recipe	61.3	2.14	13.4	10.3	14.2	200	836
55	**Chicken pie**, individual, chilled/frozen, *baked*	12 samples including chicken, chicken and ham, chicken and mushroom and chicken and vegetable pies. 10.5-25% meat	45.6	1.44	9.0	17.7	24.6	288	1202
56	**Cornish pastie**	10 samples, 5 brands	46.5	1.07	6.7	16.3	25.0	267	1117
57	-, homemade	Recipe	46.0	1.07	6.7	18.2	27.6[b]	294	1229
58	**Game pie**	Recipe	29.3	1.95	12.2	22.5	34.7[b]	381	1595
59	**Lamb samosa**	8 samples. 20-23% meat	41.3	1.65	10.3	13.4	29.1	271	1137
60	-, homemade, *baked*	Recipe	48.8	1.81	11.3	14.9	23.3[c]	267	1116
61	-, -, *deep-fried*	Recipe	42.6	1.30	8.1	31.4	16.7[d]	378	1567
62	**Pork and egg pie**	10 samples, 8 brands including prepacked	48.3	1.68	10.5	21.0	17.4	296	1234
63	**Pork pie**, individual	8 samples of 8cm pies including Melton Mowbray. 28-39% meat	37.5	1.73	10.8	25.7	23.7	363	1514
64	-, mini	10 samples of 4cm pies including buffet, Melton Mowbray and snack pies. 25-34% meat	32.1	1.70	10.6	27.8	26.3	391	1630
65	-, sliced	10 samples of own brands from the delicatessen	35.9	1.63	10.2	29.9	18.7	380	1579
66	**Sausage rolls**, puff pastry	Manufacturers' data	N	1.58	9.9	27.6	25.4	383	1596

a Includes 0.3g oligosaccharides per 100g food
b Includes 0.1g oligosaccharides per 100g food
c Includes 0.7g oligosaccharides per 100g food
d Includes 0.5g oligosaccharides per 100g food

Meat pies and pastries

No. 19-	Food	Starch g	Total sugars g	Dietary fibre Southgate method g	Dietary fibre Englyst method g	Fatty acids cis & trans Satd g	Fatty acids cis & trans Mono-unsatd g	Fatty acids cis & trans Poly-unsatd g	Total trans g	Cholesterol mg
51	**Beef pie**, chilled/frozen, *baked*	21.1	2.0	2.7	N	8.5	9.0	2.1	1.5	38
52	**Beef steak pudding**, homemade	17.5	1.0	1.0	0.9	5.4	3.9	0.5	0.4	37
53	**Bridie/Scotch pie**, individual	19.0	1.9	0.7	N	4.8	4.6	0.3	0.5	17
54	**Chicken and mushroom pie**, single crust, homemade	12.1	2.1	0.9	0.6	4.5	3.7	1.4	0.4	46
55	**Chicken pie**, individual, chilled/frozen, *baked*	23.0	1.6	2.4	0.8	7.0	7.4	2.4	1.2	32
56	**Cornish pastie**	24.1	0.9	3.4	0.9	5.9	8.4	1.2	3.4	33
57	-, homemade	25.2	2.3	2.0	1.7	6.3	7.3	3.3	0.8	36
58	**Game pie**	31.2	3.4	1.5	1.3	7.9	9.0	4.0	0.2	60
59	**Lamb samosa**	26.4	2.7	N	2.4	(1.2)	(7.7)	(4.4)	N	N
60	-, homemade, *baked*	21.0	1.6	1.8	1.7	3.5	5.3	4.8	0.4	31
61	-, -, *deep-fried*	15.0	1.2	1.3	1.2	4.7	11.2	13.5	0.3	22
62	**Pork and egg pie**	16.1	1.3	2.7	N	7.4	9.2	2.3	0.5	100
63	**Pork pie**, individual	22.7	1.0	0.9	0.9	9.7	11.0	3.2	0.4	45
64	-, mini	25.6	0.7	N	1.0	11.3	11.5	3.5	0.6	49
65	-, sliced	18.7	0	1.9	N	11.3	13.2	3.3	Tr	52
66	**Sausage rolls**, puff pastry	(24.5)	0.9	N	(1.0)	11.2	N	N	N	N

Meat pies and pastries

19-051 to 19-066

Inorganic constituents per 100g food

No. 19-	Food	Na	K	Ca	Mg	P	Fe (mg)	Cu	Zn	Cl	Mn	Se (µg)	I (µg)
51	**Beef pie**, chilled/frozen, *baked*	460	100	48	11	77	1.1	0.17	0.7	730	0.17	N	N
52	**Beef steak pudding,** homemade	340	200	40	14	150	1.3	0.06	2.4	430	0.14	4	9
53	**Bridie/Scotch pie**, individual	440	120	49	16	75	1.3	0.20	0.7	770	0.27	N	N
54	**Chicken and mushroom pie,** single crust, homemade	280	250	68	20	150	0.6	0.17	0.6	440	0.11	7	N
55	**Chicken pie**, individual, chilled/frozen, *baked*	430	140	60	15	90	0.8	0.06	0.6	710	0.23	N	N
56	**Cornish pastie**	400	140	60	14	75	1.1	0.31	0.6	720	0.20	N	3
57	-, homemade	310	200	57	14	80	1.0	0.07	1.0	490	(0.23)	(3)	N
58	**Game pie**	430	170	64	18	120	2.1	0.15	1.2	650	0.28	4	8
59	**Lamb samosa**	460	220	80	25	110	2.6	0.06	1.4	740	0.43	N	N
60	-, homemade, *baked*	210	290	50	22	130	1.6	0.10	1.8	330	0.25	2	7
61	-, -, *deep-fried*	150	200	36	16	92	1.1	0.07	1.3	240	0.18	2	7
62	**Pork and egg pie**	710	140	33	11	140	0.8	0.16	0.8	1040	0.12	N	N
63	**Pork pie**, individual	650	160	68	17	100	1.1	0.08	1.0	760	0.23	6	7
64	-, mini	680	150	60	16	110	1.2	0.10	0.8	1210	0.25	(6)	(7)
65	-, sliced	670	140	17	17	100	0.9	0.18	0.9	1180	0.20	N	N
66	**Sausage rolls**, puff pastry	600	N	N	N	N	N	N	N	N	N	N	N

Meat pies and pastries

No. 19-	Food	Retinol µg	Carotene µg	Vitamin D µg	Vitamin E mg	Thiamin mg	Ribo-flavin mg	Niacin mg	Trypt 60 mg	Vitamin B6 mg	Vitamin B12 µg	Folate µg	Panto-thenate mg	Biotin µg	Vitamin C mg
51	**Beef pie**, chilled/frozen, *baked*	Tr	Tr	N	N	0.25	0.06	1.4	N	0.18	1	15	N	N	Tr
52	**Beef steak pudding**, homemade	4	7	0.3	0.25	0.09	0.09	1.7	2.3	0.19	1	5	0.28	1	Tr
53	**Bridie/Scotch pie**, individual	Tr	Tr	N	N	0.10	0.07	1.1	N	0.10	Tr	10	N	N	Tr
54	**Chicken and mushroom pie**, single crust, homemade	57	34	(0.4)	(0.69)	0.08	0.15	4.0	2.5	0.22	Tr	8	0.72	3	Tr
55	**Chicken pie**, individual, chilled/frozen, *baked*	Tr	Tr	N	N	0.41	0.09	1.5	1.6	0.12	Tr	8	0.64	4	N
56	*Cornish pastie*	Tr	N	N	1.30	0.09	0.06	1.3	1.7	0.19	Tr	5	0.60	1	Tr
57	*-, homemade*	54	760	0.7	N	0.14	0.04	1.4	1.4	0.19	Tr	11	0.25	1	4
58	**Game pie**	610	77	(0.9)	(0.96)	0.21	0.24	3.0	2.4	0.21	3	75	0.62	12	1
59	**Lamb samosa**	Tr	59	N	0.91	0.21	0.09	2.3	2.2	0.15	1	6	0.75	2	Tr
60	*-, homemade, baked*	Tr	42	0.2	(0.18)	0.20	0.07	2.2	2.2	0.18	1	8	0.43	1	2
61	*-, -, deep-fried*	Tr	30	0.2	N	0.14	0.05	1.6	1.6	0.13	Tr	6	0.31	1	2
62	**Pork and egg pie**	Tr	Tr	N	N	0.14	0.11	1.7	N	0.15	Tr	23	N	N	N
63	**Pork pie**, individual	Tr	Tr	N	0.21	0.19	0.06	2.3	1.0	0.15	Tr	5	0.74	2	4
64	*-, mini*	Tr	Tr	N	0.22	0.20	0.06	2.0	1.3	0.13	Tr	4	0.57	2	(4)
65	*-, sliced*	Tr	Tr	N	N	0.28	0.10	1.9	N	0.17	Tr	4	N	N	N
66	**Sausage rolls**, puff pastry	N	N	N	N	N	N	N	N	N	N	N	N	N	N

Composition of food per 100g

No. 19-	Food	Description and main data sources	Water g	Total Nitrogen g	Protein g	Fat g	Carbo-hydrate g	Energy value kcal	Energy value kJ
67	**Sausage rolls**, flaky pastry, homemade	Recipe	33.0	1.57	9.8	29.3	25.0	397	1651
68	-, short pastry, homemade	Recipe	33.0	1.57	9.8	26.0	28.7	381	1588
69	**Steak and kidney/Beef pie**, individual, chilled/frozen, *baked*	16 samples including minced beef, minced beef and onion, minced beef and vegetable, steak, and steak and kidney pies. 12.5-30% meat	41.4	1.41	8.8	19.4	26.7	310	1295
70	**Steak and kidney pie**, single crust, homemade	Recipe	49.3	2.62	16.4	16.4	15.8	273	1138
71	-, double crust, homemade	Recipe	40.0	2.19	13.7	22.0	22.8	338	1412
72	**Steak and kidney pudding**, canned	10 samples, 4 brands	59.7	1.33	8.3	11.6	17.7	204	854
73	-, homemade	Recipe	58.7	1.79	11.2	10.1	19.0[a]	207	868
74	**Turkey pie**, single crust, homemade	Recipe	60.3	2.18	13.6	10.3	13.1[b]	196	822

[a] Includes 0.3g oligosaccharides per 100g food

[b] Includes 0.4g oligosaccharides per 100g food

Meat pies and pastries *continued*

No. 19-	Food	Starch g	Total sugars g	Dietary fibre Southgate method g	Dietary fibre Englyst method g	Satd g	Fatty acids cis & trans Mono-unsatd g	Fatty acids cis & trans Poly-unsatd g	Total trans g	Cholesterol mg
67	**Sausage rolls,** flaky pastry, homemade	22.7	2.3	1.3	1.4	10.9	12.6	4.7	0.8	58
68	-, short pastry, homemade	26.5	2.2	1.4	1.5	9.7	11.1	4.1	0.7	53
69	**Steak and kidney/Beef pie,** individual, chilled/frozen, *baked*	25.2	1.5	N	0.5	8.4	7.8	1.9	1.2	39
70	**Steak and kidney pie,** single crust, homemade	15.4	0.4	0.7	0.6	6.1	6.7	2.5	0.8	125
71	-, double crust, homemade	22.3	0.5	1.1	0.9	8.1	9.0	3.6	1.1	105
72	**Steak and kidney pudding,** canned	16.9	0.8	0.2	N	5.3	4.5	1.0	0.8	54
73	-, homemade	17.5	1.0	1.0	0.9	5.3	3.7	0.5	0.4	53
74	**Turkey pie,** single crust, homemade	12.2	2.5	1.3	1.0	4.5	3.7	1.5	0.4	51

Meat pies and pastries *continued*

No. Food						mg								µg	
19-	Na	K	Ca	Mg	P	Fe	Cu	Zn	Cl	Mn			Se	I
67 **Sausage rolls,** flaky pastry, homemade	730	150	98	14	150	1.1	0.09	0.8	1110	0.24			(4)	(7)
68 -, short pastry, homemade	700	150	100	14	150	1.2	0.09	0.8	1090	0.27			(4)	N
69 **Steak and kidney/Beef pie,** individual, chilled/frozen, *baked*	460	140	60	15	95	1.3	0.07	1.4	(690)	0.26			N	N
70 **Steak and kidney pie,** single crust, homemade	680	260	34	19	170	2.7	0.20	3.2	1030	0.16			38	N
71 -, double crust, homemade	620	210	46	17	150	2.3	0.18	2.5	950	0.20			30	N
72 **Steak and kidney pudding,** canned	400	110	32	9	75	1.3	0.40	1.0	550	0.12			N	N
73 -, homemade	350	190	40	14	150	1.8	0.09	2.1	450	0.15			16	9
74 **Turkey pie,** single crust, homemade	290	230	70	19	150	0.8	0.06	1.1	430	0.12			7	N

Meat pies and pastries *continued*

No. 19-	Food	Retinol µg	Carotene µg	Vitamin D µg	Vitamin E mg	Thiamin mg	Ribo-flavin mg	Niacin mg	Trypt 60 mg	Vitamin B6 mg	Vitamin B12 µg	Folate µg	Panto-thenate mg	Biotin µg	Vitamin C mg
67	**Sausage rolls,** flaky pastry, homemade	56	46	0.5	1.92	0.07	0.06	1.7	1.4	0.10	Tr	5	0.47	3	2
68	-, short pastry, homemade	47	38	0.5	1.68	0.08	0.06	1.7	1.5	0.10	Tr	6	0.46	3	2
69	**Steak and kidney/Beef pie,** individual, chilled/frozen, *baked*	5	20	0.7	1.04	0.40	(0.15)	(1.6)	(1.6)	(0.06)	(2)	(8)	0.61	3	Tr
70	**Steak and kidney pie,** single crust, homemade	69	38	1.0	1.28	0.14	0.55	3.5	3.2	0.29	4	5	1.11	9	1
71	-, double crust, homemade	87	58	1.2	1.81	0.14	0.42	2.8	2.7	0.24	3	5	0.87	7	1
72	**Steak and kidney pudding,** canned	Tr	Tr	N	N	0.04	0.28	0.1	N	0.09	3	12	N	N	Tr
73	-, homemade	13	43	0.2	0.27	0.12	0.25	1.9	2.2	0.18	2	7	0.45	2	1
74	**Turkey pie,** single crust, homemade	55	570	(0.4)	(0.62)	0.09	0.11	3.1	2.7	0.19	Tr	8	0.44	2	1

Sausages

Composition of food per 100g

No. 19-	Food	Description and main data sources	Water g	Total Nitrogen g	Protein g	Fat g	Carbohydrate g	Energy value kcal	Energy value kJ
75	**Beef sausages**, chilled, *raw*	6 samples of thick sausages. 50-55% meat	55.1	1.65	10.3	23.8	9.4	291	1206
76	-, *fried*	6 samples of thick sausages	48.1	2.18	13.6	19.7	12.5	279	1160
77	-, *grilled*	6 samples of thick sausages	48.0	2.13	13.3	19.5	13.1	278	·1157
78	**Pork sausages**, chilled, *raw*	10 samples, 7 brands of thick and thin sausages. 65-75% meat	51.5	1.89	11.8	22.7	9.2	286	1188
79	-, *fried*	16 samples	46.4	2.22	13.9	23.9	9.9	308	1279
80	-, *grilled*	16 samples	45.9	2.32	14.5	22.1	9.8	294	1221
81	frozen, *raw*	8 samples, 5 brands of thick and thin sausages. 65% meat	47.3	1.94	12.1	27.3	10.0	332	1376
82	-, *fried*	Analyses and calculation from No 81	45.3	2.21	13.8	24.8	10.0	316	1312
83	-, *grilled*	10 samples	45.1	2.37	14.8	21.2	10.5	289	1204
84	reduced fat, chilled/frozen, *raw*	7 samples, 5 brands of thin and thick sausages. 50-65% meat	62.2	2.08	13.0	10.6	8.7	180	752
85	-, -, *fried*	7 samples, 5 brands of thin and thick sausages	53.9	2.38	14.9	13.0	9.1	211	880
86	-, -, *grilled*	7 samples, 5 brands of thin and thick sausages	50.1	2.59	16.2	13.8	10.8	230	959
87	**Pork and beef sausages**, chilled, *raw*	10 samples, 7 brands	51.8	1.63	10.2	23.0	7.5	276	1144
88	-, *grilled*	10 samples	47.8	2.13	13.3	20.3	8.9	269	1120
89	frozen, *raw*	10 samples, 4 brands	44.4	1.50	9.4	28.8	11.5	340	1409
90	-, **economy**, chilled, *raw*	16 samples, 2 brands. 50-51% meat	54.6	1.63	10.2	20.5	8.9	259	1074
91	-, -, *fried*	16 samples and calculation from No 90	50.3	1.97	12.3	17.2	11.8	248	1034
92	-, -, *grilled*	16 samples and calculation from No 90	46.0	2.08	13.0	17.7	12.3	257	1073

Sausages

Composition of food per 100g

No. 19-	Food	Starch g	Total sugars g	Dietary fibre Southgate method g	Englyst method g	Fatty acids cis & trans Satd g	Mono- unsatd g	Poly- unsatd g	Total trans g	Cholesterol mg
75	**Beef sausages**, chilled, *raw*	7.9	1.5	0.7	0.7	9.5	10.9	1.8	0.6	43
76	-, *fried*	11.0	1.5	0.8	0.7	7.5	9.1	1.8	0.3	42
77	-, *grilled*	11.7	1.4	0.8	0.7	7.9	8.8	1.4	0.4	42
78	**Pork sausages**, chilled, *raw*	6.5	2.7	0.6	1.0	8.6	10.3	3.2	0.1	62
79	-, *fried*	8.4	1.6	0.7	0.7	8.5	10.3	3.5	0.1	53
80	-, *grilled*	8.3	1.5	0.7	0.7	8.0	9.6	3.0	0.1	53
81	frozen, *raw*	7.2	2.8	N	N	9.7	12.0	3.6	0.2	58
82	-, *fried*	9.5	0.5	N	N	8.8	10.9	3.3	0.2	(60)
83	-, *grilled*	10.0	0.5	N	N	7.6	9.3	2.8	0.1	(60)
84	reduced fat, chilled/frozen, *raw*	6.9	1.8	N	1.2	3.7	4.5	1.6	0.1	44
85	-, -, *fried*	8.4	0.7	N	1.4	4.2	5.6	2.3	Tr	49
86	-, -, *grilled*	9.9	0.9	N	1.5	4.9	5.9	2.1	0.1	55
87	**Pork and beef sausages**, chilled, *raw*	6.5	1.0	0.2	N	8.4	10.4	2.5	0.2	48 .
88	-, *grilled*	7.6	1.4	0.3	N	7.5	9.1	2.2	Tr	47
89	frozen, *raw*	10.7	0.8	1.3	N	11.3	12.7	2.6	0.4	91
90	-, **economy**, chilled, *raw*	5.8	3.1	N	1.1	7.5	9.0	2.7	0.1	43
91	-, -, *fried*	9.7	2.1	N	(1.2)	4.5	5.4	1.6	(0.1)	(47)
92	-, -, *grilled*	10.1	2.2	N	(1.3)	4.7	5.7	1.7	(0.1)	(51)

No. Food	Na	K	Ca	Mg	P	Fe	Cu	Zn	Cl	Mn	Se	I
						mg					µg	
19-												
75 **Beef sausages**, chilled, *raw*	920	150	60	11	140	1.1	0.08	1.5	1190	0.14	4	7
76 -, *fried*	1170	190	80	15	190	1.4	0.10	2.0	1490	0.22	(4)	(6)
77 -, *grilled*	1200	190	80	15	200	1.4	0.11	2.0	1640	0.21	4	6
78 **Pork sausages**, chilled, *raw*	880	170	95	13	190	0.9	0.07	1.1	1340	0.15	(5)	(7)
79 -, *fried*	1070	180	110	15	220	1.1	0.12	1.1	1430	0.19	6	(8)
80 -, *grilled*	1080	190	110	15	220	1.1	0.11	1.4	1660	0.20	(6)	8
81 frozen, *raw*	840	150	110	12	160	0.9	0.07	0.8	1130	0.18	(5)	(7)
82 -, *fried*	(70)	(160)	(110)	(13)	(170)	(0.9)	(0.07)	(0.8)	(1170)	(0.19)	(5)	(7)
83 -, *grilled*	(880)	(160)	(110)	(13)	(170)	(0.9)	(0.07)	(0.8)	(1180)	(0.19)	(5)	(7)
84 reduced fat, chilled/frozen, *raw*	860	170	91	14	170	0.9	0.11	1.2	1150	0.19	(6)	(7)
85 -, -, *fried*	970	210	110	16	200	1.2	0.08	1.4	1390	0.21	(6)	(8)
86 -, -, *grilled*	1180	260	130	19	230	1.3	0.08	1.7	1580	0.24	(7)	(9)
87 **Pork and beef sausages**, chilled, *raw*	780	130	30	13	300	1.0	0.23	0.8	1050	0.16	(4)	(6)
88 -, *grilled*	870	180	40	18	340	1.4	0.21	1.1	1140	0.21	(4)	(6)
89 frozen, *raw*	870	120	30	12	490	1.0	0.20	0.9	1370	0.16	(4)	(5)
90 -, **economy**, chilled, *raw*	770	130	90	9	150	0.8	0.07	0.7	1080	0.16	N	N
91 -, -, *fried*	(840)	(140)	(100)	(10)	(170)	(0.9)	(0.08)	(0.8)	(1180)	(0.18)	N	N
92 -, -, *grilled*	(920)	(160)	(110)	(11)	(180)	(1.0)	(0.08)	(0.8)	(1290)	(0.19)	N	N

No. 19-	Food	Retinol µg	Carotene µg	Vitamin D µg	Vitamin E mg	Thiamin mg	Ribo-flavin mg	Niacin mg	Trypt 60 mg	Vitamin B6 mg	Vitamin B12 µg	Folate µg	Panto-thenate mg	Biotin µg	Vitamin C mg
75	**Beef sausages**, chilled, *raw*	Tr	Tr	N	0.66	0.01	0.08	2.1	1.7	0.10	1	10	0.64	3	27[a]
76	-, *fried*	Tr	Tr	N	0.90	Tr	0.09	2.3	1.8	0.10	1	12	0.75	3	N[a]
77	-, *grilled*	Tr	Tr	N	0.67	Tr	0.11	2.5	1.8	0.11	1	7	0.64	3	N[a]
78	**Pork sausages**, chilled, *raw*	Tr	Tr	0.9	1.17	0.01	0.11	2.8	1.5	0.14	1	8	0.81	5	7
79	-, *fried*	Tr	Tr	(1.1)	0.86	0.01	0.13	3.1	2.0	0.09	1	3	0.85	5	5
80	-, *grilled*	Tr	Tr	(1.1)	0.92	Tr	0.13	3.1	2.0	0.12	1	4	0.93	5	5
81	frozen, *raw*	Tr	Tr	N	0.68	0.05	0.10	2.3	1.4	0.09	1	18	0.74	5	7
82	-, *fried*	Tr	Tr	N	N	(0.04)	(0.08)	(1.9)	(1.4)	(0.07)	(1)	(19)	(0.61)	(5)	(4)
83	-, *grilled*	Tr	Tr	N	(0.71)	(0.04)	(0.08)	(1.9)	(1.4)	(0.07)	(1)	(19)	(0.62)	(5)	(4)
84	reduced fat, chilled/frozen, *raw*	Tr	Tr	N	0.25	0.03	0.11	2.1	1.5	0.21	Tr	5	0.74	3	41
85	-, -, *fried*	Tr	Tr	N	0.28	Tr	0.12	2.2	1.8	0.12	1	5	0.96	3	32
86	-, -, *grilled*	Tr	Tr	N	0.30	Tr	0.13	2.8	2.0	0.11	1	32	1.04	3	37
87	**Pork and beef sausages**, chilled, *raw*	Tr	Tr	N	(0.64)	0.04	0.10	1.4	(1.5)	0.12	1	8	0.50	2	Tr
88	-, *grilled*	Tr	Tr	N	(0.67)	0.04	0.13	2.5	(1.8)	0.14	1	14	0.50	2	Tr
89	frozen, *raw*	Tr	Tr	N	(0.60)	0.02	0.10	1.6	(1.4)	0.12	Tr	2	0.40	3	Tr
90	-, **economy**, chilled, *raw*	Tr	Tr	N	0.24	0.01	0.08	1.7	1.5	0.11	1	5	0.71	3	N
91	-, -, *fried*	Tr	Tr	N	N	(0.01)	(0.07)	(1.5)	(1.6)	(0.14)	(1)	(6)	(0.62)	(3)	N[a]
92	-, -, *grilled*	Tr	Tr	N	0.23	(0.01)	(0.08)	(1.6)	(1.7)	(0.10)	(1)	(6)	(0.67)	(3)	N[a]

[a] Ascorbic acid is added as an antioxidant. Measurable levels may be present

Sausages *continued*

No. 19-	Food	Description and main data sources	Water g	Total Nitrogen g	Protein g	Fat g	Carbo- hydrate g	Energy value kcal	kJ
93	**Premium sausages**, chilled, *raw*	10 samples, 9 brands including Cumberland and Lincolnshire sausages. 65-90% meat	55.4	2.14	13.4	18.7	5.7	243	1011
94	-, *fried*	Analyses and calculation from No 93	50.5	2.53	15.8	20.7	6.7	275	1142
95	-, *grilled*	Analyses and calculation from No 93	49.3	2.69	16.8	22.4	6.3	292	1215

Sausages continued

Composition of food per 100g

No. 19-	Food	Starch g	Total sugars g	Dietary fibre Southgate method g	Dietary fibre Englyst method g	Fatty acids Satd g	Fatty acids cis & trans Mono-unsatd g	Fatty acids cis & trans Poly-unsatd g	Total trans g	Cholesterol mg
93	**Premium sausages**, chilled, *raw*	4.2	1.5	N	N	6.9	7.8	2.7	0.1	63
94	-, *fried*	5.7	1.0	N	N	7.7	8.5	3.0	0.2	69
95	-, *grilled*	5.4	0.9	N	N	8.2	9.4	3.3	(0.1)	72

Sausages *continued*

Inorganic constituents per 100g food

No. Food 19-	Na	K	Ca	Mg	P	mg Fe	Cu	Zn	Cl	Mn	µg Se	I
93 **Premium sausages**, chilled, *raw*	740	190	160	14	160	1.0	0.06	1.2	810	0.14	N	N
94 -, *fried*	(820)	(210)	(180)	(16)	(180)	(1.1)	(0.07)	(1.3)	(900)	(0.16)	N	N
95 -, *grilled*	(840)	(220)	(180)	(16)	(180)	(1.2)	(0.07)	(1.4)	(920)	(0.16)	N	N

Sausages *continued*

No. 19-	Food	Retinol µg	Carotene µg	Vitamin D µg	Vitamin E mg	Thiamin mg	Ribo-flavin mg	Niacin mg	Trypt 60 mg	Vitamin B6 mg	Vitamin B12 µg	Folate µg	Panto-thenate mg	Biotin µg	Vitamin C mg
93	**Premium sausages**, chilled, *raw*	Tr	Tr	N	0.90	0.05	0.11	3.0	1.3	0.16	1	7	0.84	3	14
94	-, *fried*	Tr	Tr	N	N	(0.05)	(0.10)	(2.7)	(1.5)	(0.14)	(1)	(8)	(0.74)	(3)	(8)
95	-, *grilled*	Tr	Tr	N	(0.80)	(0.05)	(0.10)	(2.7)	(1.5)	(0.14)	(1)	(8)	(0.76)	(3)	(8)

Continental style sausages

Composition of food per 100g

No. 19-	Food	Description and main data sources	Water g	Total Nitrogen g	Protein g	Fat g	Carbo-hydrate g	Energy value kcal	kJ
96	Bierwurst	Reference 7 and manufacturers' data	52.9	2.27	14.2	24.7	1.2	284	1175
97	Bratwurst	Reference 6 and manufacturers' data	56.1	2.46	15.4	20.8	3.1	260	1081
98	Cervelat	References 2,3	46.4	2.35	14.7	32.1	4.0	363	1502
99	Chorizo	Manufacturers' data	N	2.88	18.0	23.0	3.2	291	1208
100	Frankfurter	10 samples, 7 brands of continental-style frankfurters. 75-90% meat	54.2	2.17	13.6	25.4	1.1	287	1189
101	-, with bun	Recipe	39.6	1.68	10.5	11.4	33.5	270	1141
102	-, -, ketchup, fried onions and mustard	Recipe	51.2	1.12	7.0	10.2	26.1[a]	218	914
103	Garlic sausage	10 samples, 8 brands including French and German style. 65-90% meat	59.6	2.54	15.9	20.2	0.8	248	1031
104	Kabana	Reference 1 and manufacturers' data	57.3	2.67	16.7	26.6	0.6	308	1278
105	Knackwurst	Reference 4	50.1	1.90	11.9	33.7	0.3	352	1454
106	Liver sausage	10 samples, 4 brands	58.4	2.14	13.4	16.7	6.0	226	942
107	Mortadella	References 1,2,3,4 and manufacturers' data	48.3	2.27	14.2	30.1	0.8	331	1368
108	Pepperami	10 samples, 2 brands	19.9	3.57	22.3	51.1	0.6	551	2279
109	Polony	24 samples	52.0	1.50	9.4	21.1	14.2	281	1168
110	Salami	22 samples including Danish, French, German and Italian. 90-100% meat	33.7	3.34	20.9[b]	39.2[b]	0.5[b]	438	1814
111	Saveloy, unbattered, takeaway	20 samples	56.1	2.20	13.8	22.3	10.8[c]	296	1233

[a] Includes 1.2g oligosaccharides per 100g food
[c] Includes 0.6g oligosaccharides per 100g food

[b] Danish salami contains 13.4g protein, 49.7g fat, 2.2g CHO, 509kcal, 2102kJ;
French salami contains 21.0g protein, 37.4g fat, 1.9g CHO, 428kcal, 1771kJ;
German salami contains 20.7g protein, 31.5g fat, 2.6g CHO, 376kcal, 1559kJ;
Italian salami contains 23.4g protein, 30.7g fat, 0.9g CHO, 373Kcal, 1548kJ per 100g food

Continental style sausages

Composition of food per 100g

No. 19-	Food	Starch g	Total sugars g	Dietary fibre Southgate method g	Dietary fibre Englyst method g	Fatty acids cis & trans Satd g	Fatty acids cis & trans Mono-unsatd g	Fatty acids cis & trans Poly-unsatd g	Total trans g	Cholesterol mg
96	**Bierwurst**	0.8	0.4	Tr	Tr	12.3	N	3.2	N	N
97	**Bratwurst**	N	N	Tr	Tr	(8.0)	(9.8)	(2.5)	N	60
98	**Cervelat**	N	N	Tr	Tr	(10.8)	(13.2)	(2.4)	N	65
99	**Chorizo**	(0.4)	(2.8)	Tr	Tr	(9.6)	(11.0)	(2.4)	N	N
100	**Frankfurter**	Tr	1.1	0.1	0.1	9.2	11.5	3.0	0.1	76
101	–, with bun	31.9	1.6	2.8	(1.1)	3.6	4.5	1.7	Tr	24
102	–, –, ketchup, fried onions and mustard	18.1	6.8	2.5	(1.6)	2.5	3.6	3.0	Tr	13
103	**Garlic sausage**	0.8	Tr	Tr	Tr	7.3	9.1	2.4	0.1	70
104	**Kabana**	0	(0.6)	Tr	Tr	(11.4)	(13.4)	(1.3)	N	N
105	**Knackwurst**			Tr	Tr	N	N	N	N	N
106	**Liver sausage**	5.0	1.0	0.5	0.7	5.3	5.7	2.3	Tr	115
107	**Mortadella**	(0.3)	(0.5)	Tr	Tr	(11.6)	(9.5)	(3.7)	N	75
108	**Pepperami**	Tr	0.6	Tr	Tr	19.5	23.1	5.1	0.3	N
109	**Polony**	(14.2)	Tr	0.5	N	N	N	N	N	40
110	**Salami**	Tr	0.5	0.1	0.1	14.6	17.7	4.4	0.1	83
111	**Saveloy**, unbattered, takeaway	8.9	1.3	N	0.8	5.6	7.2	2.7	0.4	78

Continental style sausages

Inorganic constituents per 100g food

No. 19-	Food	Na	K	Ca	Mg	P	Fe	Cu	Zn	Cl	Mn	Se	I
						mg						μg	
96	**Bierwurst**	870	170	9	12	100	1.5	0.04	2.4	N	N	N	N
97	**Bratwurst**	480	210	45	15	150	1.3	0.09	2.3	N	0.05	N	N
98	**Cervelat**	1100	210	35	13	150	3.1	N	N	N	N	N	N
99	**Chorizo**	570	240	N	N	N	N	N	N	N	N	N	N
100	**Frankfurter**	920	170	12	11	200	1.1	0.11	1.4	1280	0.02	(8)	18
101	-, with bun	670	130	93	25	170	1.9	0.12	0.9	1010	0.32	(22)	(19)
102	-, -, ketchup, fried onions and mustard	640	230	70	21	120	1.5	0.09	0.7	890	0.28	(13)	(12)
103	**Garlic sausage**	930	240	12	14	170	0.8	0.06	1.6	1440	0.02	N	N
104	**Kabana**	880	240	12	13	190	1.8	0.06	2.3	N	0.05	N	N
105	**Knackwurst**	1200	200	28	N	150	N	N	N	N	N	N	N
106	**Liver sausage**	810	180	20	14	260	6.0	0.91	2.6	1150	0.19	N	N
107	**Mortadella**	830	220	31	15	180	2.3	0.08	3.2	N	0.05	N	N
108	**Pepperami**	1790	330	11	19	180	2.2	0.11	3.9	2660	0.18	N	N
109	**Polony**	870	120	42	13	130	1.3	0.32	1.2	1160	N	N	N[b]
110	**Salami**	1800[a]	320	11	18	170	1.3	0.12	3.0	3270	0.04	(7)	(15)[b]
111	**Saveloy**, unbattered, takeaway	1150	180	81	18	230	4.5	0.11	1.2	(1770)	0.16	N	N

[a] Iodine from erythrosine is present but largely unavailable

[b] Danish salami contains 1840mg Na; French salami contains 1700mg Na; German salami contains 1500mg Na; Italian salami contains 1335mg Na per 100g food

Continental style sausages

Vitamins per 100g food

No. 19-	Food	Retinol µg	Carotene µg	Vitamin D µg	Vitamin E mg	Thiamin mg	Ribo-flavin mg	Niacin mg	Trypt 60 mg	Vitamin B6 mg	Vitamin B12 µg	Folate µg	Panto-thenate mg	Biotin µg	Vitamin C mg
96	**Bierwurst**	N	N	N	N	0.07	0.12	3.4	N	0.17	2	3	0.33	N	16
97	**Bratwurst**	N	N	N	N	0.51	0.18	3.2	N	0.21	1	N	0.32	N	N
98	**Cervelat**	Tr	Tr	0.5	0.12	0.14	0.16	2.8	N	0.14	3	4	0.40	N	N
99	**Chorizo**	N	N	N	N	N	N	N	N	N	N	N	N	N	N
100	**Frankfurter**	Tr	Tr	N	0.63	0.32	0.15	2.8	2.2	0.12	1	3	0.75	2	N
101	-, with bun	Tr	Tr	N	0.20	0.26	0.12	1.9	2.0	0.08	Tr	34	(0.44)	(1)	N
102	-, -, ketchup, fried onions and mustard	N	34	N	0.71	0.26	0.08	1.3	1.4	0.08	Tr	31	(0.29)	1	N
103	**Garlic sausage**	Tr	Tr	N	0.42	0.50	0.18	3.9	1.5	0.24	1	4	0.96	3	N
104	**Kabana**	28	13	N	N	0.07	0.16	3.2	N	N	N	N	N	N	31
105	**Knackwurst**	N	N	N	N	N	N	N	N	N	N	N	N	N	N
106	**Liver sausage**	2600	N	(0.6)	0.10	0.36	1.16	3.7	2.4	0.25	10	36	1.50	7	Tr
107	**Mortadella**	N	N	N	N	0.16	0.16	3.5	N	0.12	2	N	N	N	N
108	**Pepperami**	Tr	Tr	N	2.05	0.27	0.16	5.5	2.2	0.27	2	1	1.19	6	N
109	**Polony**	Tr	Tr	N	0.09	0.17	0.10	1.5	1.8	0.08	Tr	4	0.50	Tr	N
110	**Salami**	Tr	Tr	N	0.23	0.60	0.23	5.6	2.8	0.36	2	3	1.66	7	N
111	**Saveloy**, unbattered, takeaway	19	Tr	N	0.45	0.14	0.09	1.9	1.9	0.06	Tr	1	0.86	4	N

Composition of food per 100g

No. 19-	Food	Description and main data sources	Water g	Total Nitrogen g	Protein g	Fat g	Carbo-hydrate g	Energy value kcal	kJ
112	**Beef slices**	10 samples including topside and wafer thin roasted beef. 100% meat	68.8	4.05	25.3	3.7	0.6	137	577
113	**Black pudding**, *raw*	8 samples, 6 brands	50.3	1.46	9.1	20.6	14.8	277	1154
114	-, *dry-fried*	8 samples, 6 brands	44.3	1.65	10.3	21.5	16.6	297	1236
115	**Brawn**	10 samples	72.0	1.99	12.4	11.5	0	153	636
116	**Chicken in crumbs**, stuffed with cheese and vegetables, chilled/frozen, *baked*	7 samples, 5 brands. Fillings include cheese and broccoli, chilli, and mushroom. 40-53% meat	57.5	2.59	16.2	13.9	10.8	230	963
117	**Chicken breast in crumbs**, chilled, *raw*	4 samples. 50-60% meat	59.7	2.93	18.3	8.3	11.4	191	801
118	-, *fried*	4 samples	53.2	2.88	18.0	12.7	14.8	242	1013
119	-, *grilled*	4 samples	53.3	2.80	17.5	10.6	14.3	219	919
120	**Chicken breast, marinated with garlic and herbs**, chilled/frozen, *baked*	7 samples. 70-95% meat	65.1	3.84	24.0	9.1	1.0[a]	182	761
121	**Chicken fingers**, baked	10 samples, 2 brands	56.2	2.00	12.5	9.5	18.5	205	860
122	**Chicken goujons**, chilled/frozen, *baked*	7 samples, 6 brands. 40-53% meat	46.5	3.10	19.4	14.0	19.6	277	1161
123	**Chicken kiev**, frozen, *baked*	5 samples, 4 brands. 45-60% meat	51.6	2.98	18.6	16.9	11.1	268	1119
124	**Chicken nuggets**, takeaway	2 samples	47.8	2.99	18.7	13.0	19.5	265	1111
125	**Chicken roll**	10 samples, 3 brands	71.3	2.74	17.1	4.8	5.2	131	552

[a] Includes 0.1g oligosaccharides per 100g food

No. Food 19-	Starch g	Total sugars g	Dietary fibre Southgate method g	Dietary fibre Englyst method g	Fatty acids Satd g	cis & trans Mono-unsatd g	Poly-unsatd g	Total trans g	Cholesterol mg
112 **Beef slices**	0	0.6	0	0	1.5	1.7	0.2	(0.1)	(53)
113 **Black pudding**, raw	14.6	0.2	N	N	7.7	9.5	2.4	0.1	43
114 -, dry-fried	16.4	0.2	0.5	(0.2)	(8.5)	(8.1)	(3.6)	N	68
115 **Brawn**	0	0	0	0	N	N	N	N	52
116 **Chicken in crumbs**, stuffed with cheese and vegetables, chilled/frozen, baked	10.1	0.7	N	0.9	4.1	4.5	4.3	0.3	48
117 **Chicken breast in crumbs**, chilled, raw	9.8	1.6	N	0.7	1.7	3.1	3.0	0.5	34
118 -, fried	14.0	0.8	N	(0.7)	2.1	5.3	4.6	0.4	(33)
119 -, grilled	13.4	0.9	N	(0.7)	2.2	3.9	3.9	0.6	(33)
120 **Chicken breast, marinated with garlic and herbs**, chilled/frozen, baked	0.3	0.6	Tr	Tr	2.9	3.9	1.8	0.2	(65)
121 **Chicken fingers**, baked	17.1	1.4	3.3	N	3.0	5.5	3.3	1.0	46
122 **Chicken goujons**, chilled/frozen, baked	18.5	1.1	N	0.7	4.0	6.1	2.8	(0.2)	N
123 **Chicken kiev**, frozen, baked	10.8	0.3	N	0.6	7.1	5.5	3.5	0.4	69
124 **Chicken nuggets**, takeaway	18.4	1.1	N	0.2	3.3	6.8	2.2	1.5	55
125 **Chicken roll**	5.2	0	0.1	Tr	1.5	2.1	0.9	0.1	40

Other meat products

19-112 to 19-125

Inorganic constituents per 100g food

No. 19-	Food	Na	K	Ca	Mg	P	Fe	Cu	Zn	Cl	Mn	Se	I
						mg						µg	
112	**Beef slices**	630	300	4	23	320	2.7	0.08	4.7	870	0.01	(8)	(7)
113	**Black pudding,** *raw*	830	95	110	14	70	10.9	0.10	0.6	1380	0.35	(5)	(4)
114	-, *dry-fried*	(940)	(110)	(120)	(16)	(80)	(12.3)	(0.11)	(0.7)	(1560)	N	6	5
115	**Brawn**	750	85	38	7	60	1.0	0.19	1.3	1110	N	N	N
116	**Chicken in crumbs,** stuffed with cheese and vegetables, chilled/frozen, *baked*	440	250	62	22	220	0.6	0.04	0.7	(680)	0.10	N	N
117	**Chicken breast in crumbs,** chilled, *raw*	420	280	21	24	180	0.1	0.06	0.5	630	0.12	N	N
118	-, *fried*	(420)	(280)	(21)	(24)	(180)	(0.1)	(0.06)	(0.5)	(620)	(0.12)	N	N
119	-, *grilled*	(400)	(270)	(20)	(23)	(170)	0.1	(0.06)	(0.5)	(600)	(0.11)	N	N
120	**Chicken breast, marinated with garlic and herbs,** chilled/frozen, *baked*	360	370	29	30	250	0.7	0.09	0.9	560	0.04	N	N
121	**Chicken fingers,** baked	640	160	29	20	220	1.1	0.15	0.8	880	0.21	N	N
122	**Chicken goujons,** chilled/frozen, *baked*	500	300	40	28	210	0.8	0.06	0.7	N	0.20	9	3
123	**Chicken kiev**, frozen, *baked*	340	280	43	24	190	0.8	0.08	0.6	500	0.12	N	N
124	**Chicken nuggets,** takeaway	510	280	25	23	210	0.6	Tr	0.5	690	0.10	N	N
125	**Chicken roll**	680	190	18	18	220	0.4	0.11	0.5	1050	0.02	N	N

Other meat products

No. 19-	Food	Retinol µg	Carotene µg	Vitamin D µg	Vitamin E mg	Thiamin mg	Ribo-flavin mg	Niacin mg	Trypt 60 mg	Vitamin B6 mg	Vitamin B12 µg	Folate µg	Panto-thenate mg	Biotin µg	Vitamin C mg
112	**Beef slices**	Tr	Tr	(0.3)	(0.03)	(0.05)	(0.25)	(4.2)	(5.2)	(0.41)	(2)	(11)	(0.43)	(2)	Tr
113	**Black pudding**, raw	N	Tr	0.6	0.16	0.09	0.07	1.2	2.0	0.09	1	2	0.41	4	0
114	-, dry-fried	41	Tr	(0.7)	0.24	0.09	0.07	1.0	2.8	0.04	1	5	0.60	2	0
115	**Brawn**	Tr	Tr	N	0.06	0.05	0.08	1.0	2.3	0.05	Tr	3	0.90	Tr	0
116	**Chicken in crumbs**, stuffed with cheese and vegetables, chilled/frozen, baked	45	92	0.5	1.71	0.14	0.10	5.6	3.2	0.38	Tr	9	1.08	4	Tr
117	**Chicken breast in crumbs**, chilled, raw	Tr	Tr	N	0.62	0.14	0.07	9.6	4.2	0.62	Tr	6	1.41	2	0
118	-, fried	Tr	Tr	N	(0.61)	(0.11)	(0.06)	7.6	(4.1)	(0.49)	Tr	(6)	(1.11)	(2)	0
119	-, grilled	Tr	Tr	N	N	(0.11)	(0.05)	(7.5)	(4.1)	(0.48)	Tr	(6)	(1.10)	(2)	0
120	**Chicken breast, marinated with garlic and herbs**, chilled/frozen, baked	Tr	Tr	(0.2)	0.94	(0.06)	(0.18)	(10.0)	(4.4)	(0.43)	Tr	(8)	(1.10)	(2)	Tr
121	**Chicken fingers**, baked	Tr	Tr	N	N	1.46	0.08	4.3	N	0.45	Tr	6	N	N	0
122	**Chicken goujons**, chilled/frozen, baked	Tr	Tr	N	N	0.14	0.04	9.4	N	0.34	Tr	21	N	N	0
123	**Chicken kiev**, frozen, baked	Tr	Tr	N	0.89	0.12	0.09	9.0	4.0	0.47	Tr	11	1.49	3	Tr
124	**Chicken nuggets**, takeaway	14	Tr	N	1.29	0.09	0.10	6.3	3.9	0.29	Tr	20	1.30	7	0
125	**Chicken roll**	Tr	Tr	N	N	0.26	0.08	6.5	N	0.34	Tr	9	0.80	2	Tr

No. 19-	Food	Description and main data sources	Water g	Total Nitrogen g	Protein g	Fat g	Carbohydrate g	Energy value kcal	kJ
126	**Chicken slices**	7 samples including smoked and wafer thin chicken breast. 80-100% meat	71.8	3.71	23.2	1.5	2.0[a]	114	482
127	**Chicken tandoori**, chilled, *reheated*	7 samples, 6 brands. 95-96% meat	56.4	4.38	27.4	10.8	2.0	214	897
128	**Corned beef**, canned	10 samples, 4 brands	59.5	4.14	25.9	10.9	1.0	205	860
129	**Doner kebab**, meat only	20 samples from assorted takeaways	42.0	3.76	23.5	31.4	0	377	1561
130	-, in pitta bread with salad	Calculated from 50% doner kebab (No 129), 22% pitta bread and 28% salad	53.7	2.27	14.2	16.2	14.0[b]	255	1065
131	**Faggots in gravy**, chilled/frozen, *reheated*	8 samples, 3 brands. 11-35% meat	69.6	1.31	8.2	7.5	12.6[c]	148	619
132	**Haggis**, *boiled*	8 samples	46.2	1.71	10.7	21.7	19.2	310	1292
133	**Ham and pork**, chopped, canned	10 samples	55.8	2.27	14.2	22.4	1.4	264	1093
134	**Lamb roast**, frozen, *cooked*	2 samples. 90% meat	58.7	3.89	24.3	13.3	Tr	217	905
135	**Luncheon meat**, canned	10 samples, 9 brands	54.4	2.06	12.9	23.8	3.6	279	1158
136	-, Chinese, *steamed*	10 samples from different shops, steamed for 15 minutes	57.1	1.90	11.9	22.3	10.5	288	1195
137	**Meat loaf**, homemade	Recipe	57.5	2.69	16.8	11.0	12.6[a]	214	894
138	-, chilled/frozen, *reheated*	10 samples, 5 brands[d]	58.8	2.78	17.4	15.8	1.2	216	900
139	**Meat spread**	10 samples of a mixture of beef and ham based spreads. 70-90% meat	64.6	2.51	15.7	13.4	2.3	192	800
140	**Minced beef in gravy**, canned	10 samples, 6 brands	70.8	1.73	10.8	11.7	3.2	161	668

[a] Includes 0.3g oligosaccharides per 100g food
[c] Includes 0.1g oligosaccharides per 100g food
[b] Includes 0.2g oligosaccharides per 100g food
[d] Edible proportion is 0.93

No. Food 19-	Starch g	Total sugars g	Dietary fibre Southgate method g	Dietary fibre Englyst method g	Fatty acids Satd g	Fatty acids cis & trans Mono- unsatd g	Fatty acids cis & trans Poly- unsatd g	Total trans g	Cholesterol mg
126 **Chicken slices**	1.5	0.2	0	0	0.4	0.7	0.3	(0)	(63)
127 **Chicken tandoori**, chilled, *reheated*	1.0	1.0	Tr	Tr	3.3	5.0	2.0	0.1	120
128 **Corned beef**, canned	0	1.0	0	0	5.7	4.3	0.3	0.7	84
129 **Doner kebab**, meat only	0	0	0	0	15.3	12.0	1.4	2.7	94
130 -, in pitta bread with salad	12.3	1.5	1.2	0.8	7.8	6.1	0.9	1.4	47
131 **Faggots in gravy,** chilled/frozen, *reheated*	10.8	1.7	N	0.2	2.5	2.9	1.0	0.1	45
132 **Haggis,** *boiled*	(19.2)	Tr	N	(0.2)	7.6	6.9	1.4	N	91
133 **Ham and pork**, chopped, canned	1.2	0.2	0.3	0.3	8.2	10.4	2.2	(0.4)	60
134 **Lamb roast**, frozen, *cooked*	Tr	Tr	0	0	6.1	5.3	0.7	(1.2)	98
135 **Luncheon meat,** canned	3.6	Tr	0.4	0.2	8.7	11.0	3.0	0.4	64
136 -, Chinese, *steamed*	N	N	Tr	Tr	N	N	N	N	N
137 **Meat loaf**, homemade	10.8	1.5	(1.0)	(0.5)	4.1	4.5	1.1	0.3	73
138 -, chilled/frozen, *reheated*	0	1.2	0.7	N	5.6	7.2	1.9	0.1	69
139 **Meat spread**	2.1	0.2	Tr	Tr	5.5	5.8	1.2	0.2	62
140 **Minced beef in gravy**, canned	2.9	0.3	Tr	Tr	5.0	5.3	0.5	0.6	45

Other meat products continued

Inorganic constituents per 100g food

No. 19-	Food	Na	K	Ca	Mg	P	Fe	Cu	Zn	Cl	Mn	Se	I
							mg					µg	
126	**Chicken slices**	780	360	13	28	350	0.4	0.08	0.8	980	0.02	(11)	(5)
127	**Chicken tandoori**, chilled, reheated	590	470	58	36	280	1.8	0.12	1.5	860	0.18	(16)	(7)
128	**Corned beef**, canned	860	140	27	15	130	2.4	0.18	5.5	1560	0.02	(8)	14
129	**Doner kebab**, meat only	860	350	23	25	210	2.1	0.11	4.0	N	0.06	6	4
130	-, in pitta bread with salad	550	260	37	20	130	1.6	0.11	2.2	N	0.17	3	3
131	**Faggots in gravy**, chilled/frozen, reheated	540	120	32	10	80	1.7	0.30	0.9	(830)	0.15	N	N
132	**Haggis**, boiled	770	170	29	36	160	4.8	0.44	1.9	1200	N	N	N
133	**Ham and pork**, chopped, canned	1050	150	52	12	210	1.6	0.18	2.4	1520	0.04	N	20
134	**Lamb roast**, frozen, cooked	610	350	7	24	270	1.9	0.12	4.5	830	0.02	4	N
135	**Luncheon meat**, canned	920	120	39	10	200	1.0	0.10	1.5	1410	0.05	(7)	N[a]
136	-, Chinese, steamed	680	180	8	13	N	1.2	0.10	2.0	N	N	N	N
137	**Meat loaf**, homemade	340	310	44	23	160	1.9	0.07	2.5	490	(0.17)	(8)	(1)
138	-, chilled/frozen, reheated	800	140	42	17	220	1.1	0.13	2.2	1170	0.09	N	N
139	**Meat spread**	810	220	15	14	140	4.7	0.13	3.3	1540	0.09	N	N
140	**Minced beef in gravy**, canned	380	150	20	11	85	1.6	0.07	1.9	580	0.06	N	N

[a] Iodine from erythrosine is present but largely unavailable

No. 19-	Food	Retinol µg	Carotene µg	Vitamin D µg	Vitamin E mg	Thiamin mg	Ribo- flavin mg	Niacin mg	Trypt 60 mg	Vitamin B6 mg	Vitamin B12 µg	Folate µg	Panto- thenate mg	Biotin µg	Vitamin C mg
126	**Chicken slices**	Tr	Tr	(0.2)	(0.24)	(0.05)	(0.18)	(9.7)	(4.2)	(0.42)	Tr	(8)	(1.06)	(2)	Tr
127	**Chicken tandoori**, chilled, *reheated*	Tr	210	(0.2)	1.49	0.12	0.19	10.2	5.8	0.61	1	16	2.25	5	2
128	**Corned beef**, canned	Tr	Tr	1.3	0.78	Tr	0.20	2.6	6.5	0.18	2	5	0.40	2	0
129	**Doner kebab**, meat only	Tr	Tr	0.6	0.56	0.11	0.25	5.8	4.9	0.20	2	7	1.10	2	0
130	-, in pitta bread with salad	Tr	91	0.3	0.47	0.14	0.14	3.4	2.9	0.14	1	16	0.60	1	2
131	**Faggots in gravy**, chilled/frozen, *reheated*	1100	55	0.5	0.33	0.10	0.56	2.0	N	0.13	6	19	N	N	Tr
132	**Haggis**, *boiled*	(1800)	Tr	(0.1)	0.41	0.16	0.35	1.5	2.0	0.07	2	8	0.50	12	Tr
133	**Ham and pork**, chopped, canned	Tr	Tr	N	0.11	0.18	0.22	1.8	2.7	0.14	1	4	0.40	2	30[a]
134	**Lamb roast**, frozen, *cooked*	Tr	Tr	N	0.24	0.15	0.27	4.5	6.0	0.22	4	4	1.07	2	0
135	**Luncheon meat**, canned	Tr	Tr	N	0.11	0.06	0.15	1.2	2.7	0.10	1	13	0.50	Tr	27[a]
136	-, Chinese, *steamed*	N	N	N	N	0.14	0.13	2.3	N	N	1	1	N	N	N
137	**Meat loaf**, homemade	14	32	0.6	(0.17)	0.34	0.13	3.8	3.2	0.20	1	(8)	(0.67)	(2)	1
138	-, chilled/frozen, *reheated*	Tr	N	N	N	1.07	0.62	2.5	N	0.24	1	14	N	N	Tr
139	**Meat spread**	Tr	Tr	N	0.49	0.07	0.19	3.4	1.8	0.13	3	6	0.75	4	0
140	**Minced beef in gravy**, canned	Tr	Tr	N	N	0.28	0.16	1.7	N	0.15	1	7	N	N	Tr

[a] Some brands contain ascorbate, range 12-60mg per 100g

No. 19-	Food	Description and main data sources	Water g	Total Nitrogen g	Protein g	Fat g	Carbo-hydrate g	Energy value kcal	Energy value kJ
141	**Minced beef pie filling**, canned	10 samples, 2 brands	73.8	1.33	8.3	8.6	4.8	129	536
142	**Pastrami**	Suppliers' data including sliced and wafer thin pastrami	N	3.10	19.4	4.3	1.8	123	518
143	**Pâté, liver**	20 samples including canned	47.6	2.02	12.6	32.7	(0.8)	348	1437
144	-, in a tube	10 samples, 6 brands	53.8	2.00	12.5	25.9	0.6	285	1180
145	-, **meat**, low fat	11 samples, assorted types; pork meat and liver based. 70-80% meat	65.0	2.88	18.0	12.0	3.0	191	798
146	**Pork haslet**	10 samples, 6 brands. 69% meat	55.0	2.18	13.6	12.6	11.4	211	880
147	**Pork roast**, frozen, *cooked*	2 samples. 88% meat	59.3	4.18	26.1	12.9	Tr	221	921
148	**Pork slices**	9 samples. 100% meat	66.4	3.73	23.3	7.6	0.4	163	684
149	**Rissoles**, savoury	10 samples	50.6	1.38	8.6	16.7	21.4	265	1107
150	**Shish kebab**, meat only	20 samples from assorted takeaways	59.4	4.64	29.0	10.0	0	206	863
151	-, in pitta bread with salad	Calculated from 37% shish kebab (No 150), 27% pitta bread and 36% salad	64.1	2.16	13.5	4.1	17.2[a]	155	656
152	**Stewed steak with gravy,** canned	10 samples, 9 brands	71.0	2.59	16.2	10.1	0.6	158	659
153	**Tongue,** canned	18 samples, lamb and ox tongue	63.9	2.56	16.0	16.5	Tr	213	883
154	**Tongue slices**	7 samples of a mixture of chilled, canned and delicatessen tongue. 90-100% meat	63.0	2.99	18.7	14.0	Tr	201	836
155	**Turkey roast**, frozen, *cooked*	2 samples. 76% meat	65.1	4.27	26.7	7.0	Tr	170	713
156	**Turkey roll**	10 samples, 8 brands	64.1	2.70	16.9	9.0	4.7	166	696

[a] Includes 0.3g oligosaccharides per 100g food

No. 19-	Food	Starch g	Total sugars g	Dietary fibre Southgate method g	Englyst method g	Fatty acids *cis & trans* Satd g	Mono-unsatd g	Poly-unsatd g	Total trans g	Cholesterol mg
141	**Minced beef pie filling**, canned	4.5	0.3	Tr	Tr	4.1	3.6	0.3	0.4	42
142	**Pastrami**	0	(1.8)	0	0	1.8	1.9	0.2	N	N
143	**Pâté, liver**	0.8	0.4	Tr	Tr	9.5	11.8	3.0	Tr	170
144	-, in a tube	0.6	Tr	Tr	Tr	7.5	8.8	2.7	0.3	145
145	-, **meat**, low fat	1.7	1.3	Tr	Tr	3.5	3.9	1.5	0.1	160
146	**Pork haslet**	11.0	0.4	Tr	Tr	4.5	5.2	2.0	0.1	48
147	**Pork roast**, frozen, *cooked*	Tr	Tr	0	0	4.5	5.6	1.8	0.1	100
148	**Pork slices**	0	0.4	0	0	2.6	3.2	1.0	Tr	(70)
149	**Rissoles**, savoury	20.3	1.1	2.4	N	7.4	7.4	0.6	0.9	23
150	**Shish kebab**, meat only	0	0	0	0	3.9	4.3	0.8	0.6	90
151	-, in pitta bread with salad	15.0	1.9	1.5	1.0	1.5	1.6	0.5	0.2	33
152	**Stewed steak with gravy**, canned	0.6	Tr	Tr	Tr	4.7	4.4	0.3	0.4	38
153	**Tongue**, canned	Tr	0	0	0	6.4	7.9	1.2	N	110
154	**Tongue slices**	0	Tr	Tr	Tr	6.0	6.4	0.9	Tr	115
155	**Turkey roast**, frozen, *cooked*	Tr	Tr	0	0	2.5	3.0	1.0	0.1	71
156	**Turkey roll**	4.7	0	0.1	N	2.7	3.8	2.0	0.1	150

Other meat products *continued*

Inorganic constituents per 100g food

No. 19-	Food	Na	K	Ca	Mg	P	Fe	Cu	Zn	Cl	Mn	Se	I
						mg						µg	
141	**Minced beef pie filling**, canned	390	140	38	12	100	1.4	0.12	1.1	540	0.10	N	N
142	**Pastrami**	920	310	N	N	N	N	N	N	N	N	N	N
143	**Pâté, liver**	750	150	16	11	450	7.4	0.46	2.8	880	0.16	N	N
144	-, in a tube	830	170	17	11	520	6.0	0.67	2.7	1030	0.12	N	N
145	-, **meat**, low fat	710	190	14	14	240	6.4	0.46	2.7	1180	0.16	N	N
146	**Pork haslet**	960	220	90	16	210	1.9	0.12	1.5	1480	0.19	N	N
147	**Pork roast**, frozen, *cooked*	570	360	10	25	280	1.5	0.15	3.7	780	0.02	N	N
148	**Pork slices**	770	340	6	25	370	0.8	0.13	2.2	1010	0.02	(15)	(2)
149	**Rissoles**, savoury	520	170	55	17	120	1.5	0.33	1.2	740	0.25	N	N
150	**Shish kebab**, meat only	510	420	7	29	250	2.6	0.14	6.1	N	0.03	4	6
151	-, in pitta bread with salad	330	260	34	19	130	1.6	0.12	2.5	N	0.19	2	3
152	**Stewed steak with gravy**, canned	340	200	11	15	120	2.1	0.18	3.9	510	0.02	N	N
153	**Tongue**, canned	1050	97	32	14	140	2.5	0.29	2.3	1430	N	N	N
154	**Tongue slices**	1000	140	10	16	260	2.6	0.18	3.0	1190	0.02	8	11
155	**Turkey roast**, frozen, *cooked*	530	300	7	25	300	0.5	0.04	1.4	580	Tr	15	N
156	**Turkey roll**	690	180	15	17	200	0.8	0.11	1.5	1110	0.04	N	N

Other meat products *continued*

No. 19-	Food	Retinol µg	Carotene µg	Vitamin D µg	Vitamin E mg	Thiamin mg	Ribo-flavin mg	Niacin mg	Trypt 60 mg	Vitamin B6 mg	Vitamin B12 µg	Folate µg	Panto-thenate mg	Biotin µg	Vitamin C mg
141	**Minced beef pie filling**, canned	Tr	Tr	N	N	0.04	0.04	0.8	N	0.10	1	3	7.00	N	Tr
142	**Pastrami**	N	N	N	N	N	N	N	N	N	N	N	N	N	N
143	**Pâté, liver**	7300	130	1.2	N	0.10	1.17	1.9	2.8	0.25	8	99	2.10	14	N
144	-, in a tube	(7300)	(130)	(1.2)	N	0.12	0.88	4.2	2.8	0.23	6	19	1.80	7	N
145	-, **meat**, low fat	5930	N	N	0.77	0.46	1.12	7.1	2.2	0.35	12	31	2.68	27	18
146	**Pork haslet**	Tr	Tr	N	0.64	0.14	0.19	3.6	2.2	0.19	3	11	0.89	4	Tr
147	**Pork roast**, frozen, *cooked*	Tr	Tr	0.3a	0.11	0.63	0.25	4.8	6.7	0.30	2	4	1.62	4	0
148	**Pork slices**	Tr	Tr	(0.4)	(0.01)	(0.52)	(0.18)	(6.9)	(4.7)	(0.35)	(1)	(3)	(2.05)	(4)	Tr
149	**Rissoles**, savoury	Tr	Tr	N	N	0.06	0.07	1.3	N	0.14	1	6	N	N	Tr
150	**Shish kebab**, meat only	Tr	Tr	0.6	0.67	0.14	0.28	7.0	6.0	0.26	3	9	1.40	3	0
151	-, in pitta bread with salad	Tr	120	0.2	0.50	0.16	0.12	3.2	2.8	0.14	1	20	0.58	1	3
152	**Stewed steak with gravy,** canned	Tr	Tr	N	0.59	0.02	0.16	2.3	2.8	0.29	2	6	0.30	1	Tr
153	**Tongue,** canned	Tr	Tr	Tr	0.26	0.04	0.39	2.5	3.8	0.04	5	2	0.40	2	0
154	**Tongue slices**	Tr	Tr	N	N	0.03	0.18	2.0	N	0.12	5	4	N	N	0
155	**Turkey roast**, frozen, *cooked*	Tr	Tr	0.1a	Tr	0.06	0.13	9.3	6.5	0.41	Tr	3	0.63	2	0
156	**Turkey roll**	Tr	Tr	N	N	0.05	0.08	5.2	N	0.25	1	5	0.40	2	Tr

a Contribution from 25-hydroxycholecalciferol not included

Other meat products continued

Composition of food per 100g

No. Food 19-	Description and main data sources	Water g	Total Nitrogen g	Protein g	Fat g	Carbo-hydrate g	Energy value kcal	kJ
157 **Turkey slices**	13 samples including honey roast and wafer thin turkey breast. 78-100% meat	72.6	3.68	23.0	1.9	1.2[a]	114	481
158 **Turkey steaks in crumbs**, frozen, *grilled*	4 samples, 3 brands. 38-49% meat	45.9	2.82	17.6	17.1	18.9	295	1234
159 **White pudding**	6 samples	22.8	1.12	7.0	31.8	36.3	450	1876

[a] Includes 0.4g oligosaccharides per 100g food

No. Food	Starch	Total sugars	Dietary fibre		Fatty acids *cis & trans*			Total trans	Cholesterol
19-			Southgate method	Englyst method	Satd	Mono- unsatd	Poly- unsatd		
	g	g	g	g	g	g	g	g	mg
157 **Turkey slices**	0.4	0.4	0	0	0.6	0.7	0.4	(0)	(62)
158 **Turkey steaks in crumbs,** *frozen, grilled*	16.5	2.4	N	0.6	4.2	6.8	5.1	0.2	38
159 **White pudding**	(36.3)	Tr	3.1	N	N	N	N	N	22

Other meat products *continued*

Inorganic constituents per 100g food

No. Food						mg							µg	
19-	Na	K	Ca	Mg	P	Fe	Cu	Zn	Cl	Mn		Se	I
157 Turkey slices	750	330	6	25	310	0.4	0.13	1.1	1030	0.01		(11)	(5)
158 Turkey steaks in crumbs, frozen, *grilled*	700	240	63	21	190	1.1	0.08	1.1	940	0.17		N	6
159 White pudding	370	190	38	61	230	2.1	0.43	1.6	600	N		N	N

Other meat products *continued*

Vitamins per 100g food

No. Food 19-	Retinol μg	Carotene μg	Vitamin D μg	Vitamin E mg	Thiamin mg	Ribo- flavin mg	Niacin mg	Trypt 60 mg	Vitamin B6 mg	Vitamin B12 μg	Folate μg	Panto- thenate mg	Biotin μg	Vitamin C mg
157 **Turkey slices**	Tr	Tr	(0.2)	(0.24)	(0.05)	(0.18)	(9.6)	(4.2)	(0.41)	Tr	(8)	(1.05)	(2)	Tr
158 **Turkey steaks in crumbs,** *frozen, grilled*	12	Tr	N	2.08	2.35	0.09	5.2	3.1	0.16	Tr	13	0.80	3	0
159 **White pudding**	Tr	Tr	N	1.00	0.26	0.08	0.5	1.3	0.06	1	6	0.80	18	0

No. 19-	Food	Description and main data sources	Water g	Total Nitrogen g	Protein g	Fat g	Carbo-hydrate g	Energy value kcal	kJ
160	**Beef and spinach curry**	Recipe	76.5	1.70	10.6	5.6	3.1[a]	104	437
161	**Beef bourguignonne**	Recipe	(70.8)	2.24	14.0	6.7	2.5[b]	126	526
162	-, made with lean beef	Recipe	(73.0)	2.29	14.3	4.7	2.6[c]	109	459
163	**Beef casserole**, canned	Manufacturers' data, 5 brands	N	1.10	6.9	2.7	6.9	78	328
164	-, made with canned cook-in sauce	Recipe	71.6	2.42	15.1	6.5	4.5	136	569
165	**Beef chow mein**, retail, *reheated*	12 samples from different shops. Noodles with beef and vegetables in sauce	71.7	1.07	6.7	6.0	14.7	136	571
166	**Beef curry**	Recipe	65.4	2.18	13.6	16.0	1.7[d]	205	850
167	-, reduced fat	Recipe	70.9	2.27	14.2	9.6	1.8[d]	150	625
168	-, canned	Manufacturers' data, 11 brands including beef curry and vegetables	N	1.46	9.1	5.9	6.1	112	471
169	-, chilled/frozen, *reheated*	6 samples, 3 brands	69.5	2.16	13.5	6.6	6.3	137	575
170	-, -, with rice	Calculated from 57% beef curry (No 169) and 43% boiled white rice	69.7	1.38	8.6	3.9	16.4	131	551
171	**Beef enchiladas**	Recipe	62.6	1.70	10.6	7.8	12.0[b]	158	661
172	**Beef in sauce with vegetables**, chilled/frozen, *reheated*	8 samples, 5 brands including braised steak, beef bourguignonne and goulash	74.4	2.05	12.8	4.6	4.8[b]	111	465
173	**Beef kheema**	Recipe	70.3	1.82	11.4	11.1	3.4[e]	158	659
174	**Beef olives**	Recipe	69.3	2.22	13.9	12.9	1.7[f]	178	741
175	**Beef stew**	Recipe	75.6	1.94	12.1	5.1	5.0[c]	113	474

[a] Includes 0.5g oligosaccharides per 100g food
[c] Includes 0.1g oligosaccharides per 100g food
[e] Includes 0.4g oligosaccharides per 100g food
[b] Includes 0.3g oligosaccharides per 100g food
[d] Includes 0.2g oligosaccharides per 100g food
[f] Includes 0.7g oligosaccharides per 100g food

Meat dishes

No. 19-	Food	Starch g	Total sugars g	Dietary fibre Southgate method g	Dietary fibre Englyst method g	Fatty acids Satd g	cis & trans Mono-unsatd g	cis & trans Poly-unsatd g	Total trans g	Cholesterol mg
160	**Beef and spinach curry**	0.2	2.4	(1.2)	0.9	1.8	2.3	1.0	0.2	27
161	**Beef bourguignonne**	1.4	0.8	0.6	0.4	2.1	2.7	1.2	0.1	42
162	-, made with lean beef	1.4	0.8	0.6	0.4	1.3	1.8	1.1	0.1	40
163	**Beef casserole**, canned	(5.7)	1.2	N	(1.0)	1.4	N	N	N	N
164	-, made with canned cook-in sauce	1.8	2.7	N	N	2.7	2.9	0.4	0.3	44
165	**Beef chow mein**, retail, *reheated*	12.3	2.4	N	N	1.3	3.1	1.4	N	N
166	**Beef curry**	0.2	1.3	0.4	0.3	3.4	6.2	5.3	0.2	38
167	-, reduced fat	0.2	1.4	0.4	0.4	2.8	3.9	2.1	0.3	40
168	-, canned	(3.8)	2.3	N	0.8	1.4	N	N	N	N
169	-, chilled/frozen, *reheated*	1.8	4.5	N	1.2	3.1	2.5	0.6	N	32
170	-, with rice	13.8	2.6	0.3	0.8	1.8	1.4	0.3	N	18
171	**Beef enchiladas**	9.2	2.5	1.5	1.3	3.2	3.0	0.7	0.3	23
172	**Beef in sauce with vegetables**, chilled/frozen, *reheated*	1.5	3.0	N	0.7	N	N	N	N	N
173	**Beef kheema**	1.1	1.6	1.7	(1.3)	3.9	4.5	1.7	0.4	30
174	**Beef olives**	1.3	0.3	0.1	0.1	4.5	5.5	1.8	0.2	37
175	**Beef stew**	2.5	2.1	0.7	0.7	1.5	2.0	0.9	0.1	35

Meat dishes

Inorganic constituents per 100g food

No. 19-	Food	Na	K	Ca	Mg	P	Fe	Cu	Zn	Cl	Mn	Se	I
						mg						µg	
160	**Beef and spinach curry**	240	370	52	29	110	2.0	0.06	2.8	340	0.21	3	9
161	**Beef bourguignonne**	360	320	15	18	130	1.7	0.19	3.2	380	0.08	6	N
162	-, made with lean beef	380	340	15	19	140	1.8	0.19	3.4	430	0.08	6	N
163	**Beef casserole**, canned	300	N	N	N	N	N	N	N	N	N	N	N
164	-, made with canned cook-in sauce	560	280	7	16	140	1.2	0.02	4.0	380	0.04	N	N
165	**Beef chow mein**, retail, *reheated*	590	N	N	N	N	1.3	N	N	910	N	N	N
166	**Beef curry**	180	320	22	23	130	1.5	0.05	3.6	250	0.10	4	12
167	-, reduced fat	180	330	23	24	130	1.6	0.05	3.8	270	0.11	5	12
168	-, canned	390	N	N	N	N	N	N	N	N	N	N	N
169	-, chilled/frozen, *reheated*	450	340	N	N	N	N	N	N	690	N	N	N
170	-, -, with rice	260	210	Tr	2	15	0.1	0.03	0.2	400	0.13	2	2
171	**Beef enchiladas**	320	240	57	17	100	1.1	0.10	1.6	440	0.16	(4)	N
172	**Beef in sauce with vegetables,** chilled/frozen, *reheated*	360	190	20	13	95	1.5	0.10	3.4	N	0.08	N	N
173	**Beef kheema**	240	260	23	20	110	1.4	0.04	2.2	360	0.15	4	6
174	**Beef olives**	590	240	11	16	130	1.2	0.05	2.1	590	0.04	5	7
175	**Beef stew**	390	240	17	13	110	1.3	0.04	2.8	340	0.06	4	8

No. 19-	Food	Retinol µg	Carotene µg	Vitamin D µg	Vitamin E mg	Thiamin mg	Ribo-flavin mg	Niacin mg	Trypt 60 mg	Vitamin B6 mg	Vitamin B12 µg	Folate µg	Panto-thenate mg	Biotin µg	Vitamin C mg
160	**Beef and spinach curry**	Tr	635	0.2	(0.64)	0.08	0.12	1.9	2.1	0.24	1	27	(0.32)	(1)	6
161	**Beef bourguignonne**	Tr	13	0.4	N	0.08	0.17	2.5	2.7	0.25	1	7	0.62	3	Tr
162	-, made with lean beef	Tr	13	0.5	N	0.09	0.17	2.7	2.8	0.27	1	6	0.63	3	1
163	**Beef casserole**, canned	N	N	N	N	N	N	N	N	N	0	N	N	N	N
164	-, made with canned cook-in sauce	Tr	N	0.3	N	0.04	0.16	2.2	3.1	0.25	1	19	0.33	N	Tr
165	**Beef chow mein**, retail, *reheated*	Tr	Tr	N	(0.43)	0.03	0.03	N	1.1	N	Tr	N	N	N	Tr
166	**Beef curry**	Tr	80	0.3	(0.37)	0.06	0.15	2.2	2.7	0.25	1	19	0.35	1	2
167	-, reduced fat	Tr	84	0.3	(0.38)	0.06	0.15	2.3	2.9	0.26	1	20	0.37	1	2
168	-, canned	N	N	N	N	N	N	N	N	N	N	N	N	N	N
169	-, chilled/frozen, *reheated*	Tr	Tr	N	0.62	0.05	0.16	2.4	1.6	0.20	N	N	0.71	3	Tr
170	-, -, with rice	Tr	Tr	N	0.35	0.03	0.10	1.5	1.1	0.14	N	N	0.49	2	Tr
171	**Beef enchiladas**	11	105	0.2	(0.51)	0.08	0.06	2.0	1.8	0.18	1	9	0.25	N	6
172	**Beef in sauce with vegetables**, chilled/frozen, *reheated*	Tr	Tr	N	N	0.05	0.06	1.8	N	0.23	1	12	N	N	Tr
173	**Beef kheema**	Tr	140	0.2	(0.33)	0.10	0.07	2.8	2.0	0.20	1	13	0.26	1	4
174	**Beef olives**	Tr	8	0.4	(0.08)	0.11	0.09	2.7	2.7	0.25	1	6	0.39	1	Tr
175	**Beef stew**	Tr	1245	0.4	(0.20)	0.06	0.11	1.8	2.3	0.22	1	4	0.32	1	1

Meat dishes *continued*

Composition of food per 100g

No. 19-	Food	Description and main data sources	Water g	Total Nitrogen g	Protein g	Fat g	Carbo-hydrate g	Energy value kcal	Energy value kJ
176	**Beef stew**, made with lean beef	Recipe	77.3	1.98	12.4	3.6	5.0a	101	424
177	**Beef stew and dumplings**	Recipe	62.1	1.62	10.1	10.3	15.7b	192	804
178	-, canned	Manufacturers' data, 3 brands	N	1.09	6.8	3.7	10.0	98	413
179	-, retail, *cooked*	Calculated from 68% beef stew and 32% dumplings	64.6	1.45	9.0c	7.9c	16.2c	89c	375c
180	**Beef, stir-fried with green peppers**	Recipe	70.5	1.89	11.8	8.5	6.0	146	611
181	**Beef Stroganoff**	Recipe	68.6	2.46	15.4	11.2	2.5d	171	716
182	**Beef Wellington**	Recipe	41.0	2.58	16.1	24.9	16.5	350	1459
183	**Bolognese sauce**	Recipe	(70.5)	1.90	11.9	10.4	2.6e	151	629
184	**Cannelloni**, chilled/frozen, *reheated*	10 samples, 4 brands	73.1	1.02	6.4	5.0	13.5	121	510
185	**Carbonnade of beef**	Recipe	72.2	2.05	12.8	6.8	5.3f	132	554
186	**Chicken chasseur**	Recipe	78.6	2.05	12.8	1.9	2.6b	78	331
187	-, *weighed with bone*	Recipe e	73.5	2.02	12.6	1.7	2.3e	74	314
188	**Chicken curry**, chilled/frozen, *reheated*	7 samples, 5 brands; korma and masala varieties	68.7	1.94	12.1	8.9	5.4	149	621
189	-, with rice	Calculated from 55% chicken curry (No 188) and 45% boiled white rice	69.2	1.22	7.6	5.0	16.3	137	575
190	-, made with canned curry sauce	Recipe	65.9	3.20	20.0	6.1	4.4	151	635
191	**Chicken fricassée**	Recipe	77.2	1.68	10.5	6.2	2.6	107	448
192	-, reduced fat	Recipe	78.6	1.73	10.8	4.1	2.9	91	382

a Includes 0.4g oligosaccharides per 100g food
c Retail beef stew contains 9.3g protein, 3.8g fat, 7.7g carbohydrate, 100kcal, 422kJ per 100g food
e Includes 0.2g oligosaccharides per 100g food
b Includes 0.3g oligosaccharides per 100g food
d Includes 0.5g oligosaccharides per 100g food
f Includes 0.7g oligosaccharides per 100g food

Meat dishes *continued*

No. 19-	Food	Starch g	Total sugars g	Dietary fibre Southgate method g	Dietary fibre Englyst method g	Fatty acids Satd g	cis & trans Mono-unsatd g	Poly-unsatd g	Total trans g	Cholesterol mg
176	**Beef stew**, made with lean beef	2.5	2.1	0.7	0.7	0.9	1.3	0.9	0.1	34
177	**Beef stew and dumplings**	13.7	1.7	1.2	(1.0)	4.8	3.7	0.9	0.4	32
178	-, canned	(8.8)	1.2	N	(0.7)	N	N	N	N	N
179	-, retail, *cooked*	14.8	1.5	N	0.7	4.1	2.8	0.4	0.3	20
180	**Beef, stir-fried with green peppers**	2.1	3.6	1.0	0.8	2.7	3.4	1.9	0.2	32
181	**Beef Stroganoff**	Tr	2.0	0.8	0.5	6.2	3.7	0.6	0.2	61
182	**Beef Wellington**	16.1	0.4	0.8	0.7	8.7	9.8	4.1	0.9	82
183	**Bolognese sauce**	0.2	2.2	0.6	0.6	4.2	4.4	0.8	0.5	34
184	**Cannelloni**, chilled/frozen, *reheated*	8.4	5.1	N	1.2	2.0	2.0	0.6	0.3	12
185	**Carbonnade of beef**	1.9	2.7	0.5	0.5	2.2	2.8	1.2	0.1	39
186	**Chicken chasseur**	1.0	1.3	0.4	0.3	0.3	0.6	0.7	Tr	36
187	-, *weighed with bone*	0.9	1.2	0.4	0.3	0.3	0.6	0.6	Tr	35
188	**Chicken curry**, chilled/frozen, *reheated*	1.0	4.4	N	1.3	4.0	2.9	1.5	N	51
189	-, with rice	13.9	2.4	0.4	0.8	2.2	1.6	0.8	N	28
190	-, made with canned curry sauce	2.1	2.3	N	N	N	N	N	Tr	56
191	**Chicken fricassée**	1.3	1.3	0.6	0.5	2.4	2.1	1.2	Tr	34
192	-, reduced fat	1.3	1.6	(0.6)	(0.5)	1.0	1.6	1.1	Tr	29

Meat dishes *continued*

Inorganic constituents per 100g food

No. 19-	Food	Na	K	Ca	Mg	P	Fe	Cu	Zn	Cl	Mn	Se	I
						mg						µg	
176	**Beef stew**, made with lean beef	400	250	17	13	110	1.3	0.04	3.0	340	0.06	4	7
177	**Beef stew and dumplings**	380	190	64	12	140	1.2	0.05	2.1	310	0.13	N	8
178	-, canned	380	N	N	N	N	N	N	N	N	N	N	N
179	-, retail, *cooked*	670	190	43	20	130	0.7	0.10	0.5	890	0.24	2	3
180	**Beef, stir-fried with green peppers**	270	270	16	19	120	1.5	(0.04)	1.9	N	0.05	N	N
181	**Beef Stroganoff**	190	340	28	20	170	1.6	0.16	2.1	300	0.06	(7)	7
182	**Beef Wellington**	430	250	36	19	210	2.7	0.11	2.1	620	0.16	5	7
183	**Bolognese sauce**	250	320	17	18	110	1.2	0.08	2.3	310	0.11	4	(7)
184	**Cannelloni**, chilled/frozen, *reheated*	460	180	65	20	90	1.2	0.16	0.7	820	0.18	N	N
185	**Carbonnade of beef**	330	250	17	15	120	1.2	0.05	2.8	390	0.06	(4)	N
186	**Chicken chasseur**	230	270	13	19	130	0.6	0.13	0.5	240	0.07	7	4
187	-, *weighed with bone*	210	260	11	18	120	0.6	0.12	0.4	220	0.06	7	4
188	**Chicken curry**, chilled/frozen, *reheated*	450	300	N	N	N	N	N	N	640	N	N	N
189	-, with rice	250	180	N	N	N	N	N	N	360	N	N	N
190	-, made with canned curry sauce	650	400	22	34	190	1.1	0.07	0.7	530	0.13	N	N
191	**Chicken fricassée**	360	240	21	16	120	0.6	0.10	0.6	470	0.07	6	N
192	-, reduced fat	360	260	29	17	120	0.6	(0.10)	0.6	470	(0.07)	(6)	N

No. 19-	Food	Retinol µg	Carotene µg	Vitamin D µg	Vitamin E mg	Thiamin mg	Ribo-flavin mg	Niacin mg	Trypt 60 mg	Vitamin B6 mg	Vitamin B12 µg	Folate µg	Panto-thenate mg	Biotin µg	Vitamin C mg
176	**Beef stew**, made with lean beef	Tr	1245	0.4	0.20	0.06	0.11	1.9	2.4	0.23	1	4	0.32	1	1
177	**Beef stew and dumplings**	Tr	910	0.3	(0.27)	0.08	0.08	1.5	1.9	0.18	1	4	0.27	1	1
178	-, canned	N	N	N	N	N	N	N	N	N	N	N	N	N	N
179	-, retail, *cooked*	3	465	N	0.41	0.06	0.08	1.5	1.8	0.17	1	9	0.34	1	8
180	**Beef, stir-fried with green peppers**	Tr	(125)	0.2	(0.29)	0.05	0.11	2.3	2.5	0.38	1	13	0.31	N	27
181	**Beef Stroganoff**	84	32	0.3	(0.24)	0.11	0.23	3.1	3.2	0.32	1	11	0.94	3	1
182	**Beef Wellington**	1100	59	1.0	(1.25)	0.13	0.27	2.7	3.3	0.27	2	13	0.78	3	Tr
183	**Bolognese sauce**	Tr	480	0.3	(0.61)	0.06	0.07	3.0	2.1	0.23	1	8	0.33	1	3
184	**Cannelloni**, chilled/frozen, *reheated*	N	N	N	N	0.10	0.13	1.3	N	0.29	Tr	13	N	N	Tr
185	**Carbonnade of beef**	Tr	5	0.4	N	0.10	0.12	2.1	2.5	0.25	1	5	2.65	1	1
186	**Chicken chasseur**	Tr	10	0.1	(0.14)	0.08	0.08	4.7	2.5	0.24	Tr	7	0.67	2	Tr
187	-, *weighed with bone*	Tr	9	0.1	(0.13)	0.08	0.08	4.6	2.4	0.24	Tr	6	0.64	2	Tr
188	**Chicken curry**, chilled/frozen, *reheated*	N	370	N	1.30	0.20	0.14	3.8	2.3	0.24	N	N	1.02	3	1
189	-, with rice	N	205	N	0.72	0.11	0.08	2.2	1.5	0.15	N	N	0.65	2	1
190	-, made with canned curry sauce	Tr	N	0.1	N	0.09	0.10	6.8	3.8	0.33	Tr	N	N	N	Tr
191	**Chicken fricassée**	31	N	0.1	N	0.14	0.11	4.5	2.0	(0.29)	Tr	13	0.73	N	3
192	-, reduced fat	1	N	0.1	(0.11)	0.12	0.09	3.6	2.1	0.23	Tr	(7)	(0.60)	N	2

No. 19-	Food	Description and main data sources	Water g	Total Nitrogen g	Protein g	Fat g	Carbohydrate g	Energy value kcal	Energy value kJ
193	**Chicken in sauce with vegetables,** chilled/frozen, *reheated*	10 samples including chicken, tomato and mushroom casserole, and chicken creole	75.3	2.13	13.3	5.1	4.5[a]	116	487
194	**Chicken in white sauce,** canned	10 samples, 4 brands	73.6	2.29	14.3	8.3	2.5	141	590
195	-, made with whole milk	Recipe	69.1	2.70	16.9	8.7	4.9	164	688
196	-, made with semi-skimmed milk	Recipe	70.1	2.70	16.9	7.6	5.0	155	649
197	**Chicken korma**	Recipe	71.4	2.43	15.2	5.8	4.6[b]	130	547
198	**Chicken risotto**	Recipe	58.6	1.63	10.2	3.3	27.9[c]	182	770
199	**Chicken, stir-fried with mushrooms and cashew nuts**	Recipe	(65.4)	3.15	19.7	6.5	(4.5)[d]	154	647
200	**Chicken, stir-fried with peppers in black bean sauce**	Recipe	74.0	2.80	17.5	3.5	2.8[c]	112	472
201	**Chicken, stir-fried with rice and vegetables,** frozen, *reheated*	6 samples. 10-13% meat	67.9	1.04	6.5	4.6	17.1[e]	132	554
202	**Chicken vindaloo**	Recipe	48.3	2.93	18.3	15.1	2.5[b]	218	910
203	-, reduced fat	Recipe	(55.6)	3.01	18.8	4.1	2.6[b]	122	513
204	**Chicken wings,** marinated, chilled/frozen, *barbecued*	4 samples including American and Chinese style and hot and spicy wings	50.5	4.38	27.4	16.6	4.1	274	1146
205	-, *weighed with bone*	Calculated from No 204[d]	32.8	2.85	17.8	10.8	2.7	179	745
206	**Chilli con carne**	Recipe	68.7	1.70	10.6	7.9	4.5[b]	130	545

[a] Includes 0.2g oligosaccharides per 100g food
[b] Includes 0.4g oligosaccharides per 100g food
[c] Includes 0.3g oligosaccharides per 100g food
[d] Edible proportion is 0.65
[e] Includes 0.6g oligosaccharides per 100g food

Meat dishes *continued*

| No. 19- | Food | Starch g | Total sugars g | Dietary fibre | | Fatty acids *cis & trans* | | | Total trans g | Cholesterol mg |
				Southgate method g	Englyst method g	Satd g	Mono-unsatd g	Poly-unsatd g		
193	**Chicken in sauce with vegetables,** chilled/frozen, *reheated*	1.5	2.8	N	0.3	N	N	N	N	N
194	**Chicken in white sauce,** canned	2.5	0	Tr	Tr	2.3	3.9	1.6	Tr	49
195	-, made with whole milk	2.5	2.4	0.1	0.1	3.1	3.4	1.6	0.4	69
196	-, made with semi-skimmed milk	2.5	2.5	0.1	0.1	2.5	3.2	1.5	0.4	66
197	**Chicken korma**	1.4	2.8	0.5	0.4	1.7	1.9	1.8	Tr	44
198	**Chicken risotto**	N	N	N	N	1.5	0.7	0.1	0.1	25
199	**Chicken, stir-fried with mushrooms and cashew nuts**	(2.5)	(2.0)	(0.6)	0.7	1.2	3.1	1.9	Tr	51
200	**Chicken, stir-fried with peppers in black bean sauce**	0.1	2.4	0.6	0.8	0.5	1.2	1.4	Tr	48
201	**Chicken, stir-fried with rice and vegetables,** frozen, *reheated*	12.6	3.9	N	1.3	N	N	N	N	N
202	**Chicken vindaloo**	0.4	1.7	0.2	0.3	1.8	5.7	6.6	Tr	69
203	-, reduced fat	0.4	1.8	0.2	0.3	0.7	1.6	1.5	Tr	74
204	**Chicken wings,** marinated, chilled/frozen, *barbecued*	0.5	3.6	Tr	Tr	4.6	7.5	3.3	(0.2)	(120)
205	-, *weighed with bone*	0.3	2.3	Tr	Tr	3.0	4.9	2.1	(0.1)	(78)
206	**Chilli con carne**	1.3	2.8	1.4	1.1	3.0	3.2	0.8	0.3	24

Meat dishes *continued*

Inorganic constituents per 100g food

No. 19-	Food	Na	K	Ca	Mg	P	Fe	Cu	Zn	Cl	Mn	Se	I
						mg						µg	
193	Chicken in sauce with vegetables, chilled/frozen, reheated	320	340	19	18	120	0.7	0.11	0.7	(500)	0.09	N	N
194	Chicken in white sauce, canned	370	80	13	11	70	0.6	0.07	0.9	470	0.03	N	N
195	-, made with whole milk	230	250	66	21	170	0.5	0.06	1.0	330	0.01	9	12
196	-, made with semi-skimmed milk	230	260	68	21	170	0.5	0.06	1.0	330	0.01	9	12
197	Chicken korma	170	330	61	27	170	0.8	0.06	0.7	260	0.09	(7)	(11)
198	Chicken risotto	420	230	27	15	120	0.8	0.12	1.0	330	0.39	N	N
199	Chicken, stir-fried with mushrooms and cashew nuts	300	410	10	47	220	1.1	0.26	1.0	N	0.17	(12)	(6)
200	Chicken, stir-fried with peppers in black bean sauce	300	320	17	28	170	1.1	0.07	0.6	490	0.09	8	5
201	Chicken, stir-fried with rice and vegetables, frozen, reheated	410	180	22	13	95	1.1	0.10	2.0	870	0.08	N	N
202	Chicken vindaloo	210	400	41	37	150	2.7	0.09	1.2	310	0.23	10	7
203	-, reduced fat	230	370	12	25	140	0.8	0.04	1.1	340	0.07	(11)	6
204	Chicken wings, marinated, chilled/frozen, barbecued	390	350	42	27	200	1.3	0.09	1.6	610	0.15	(17)	(6)
205	-, weighed with bone	250	230	27	18	130	0.9	0.06	1.1	400	0.10	(11)	(4)
206	Chilli con carne	310	270	20	16	91	1.1	0.10	1.7	410	0.11	(3)	N

Meat dishes continued

No. 19-	Food	Retinol µg	Carotene µg	Vitamin D µg	Vitamin E mg	Thiamin mg	Ribo-flavin mg	Niacin mg	Trypt 60 mg	Vitamin B6 mg	Vitamin B12 µg	Folate µg	Panto-thenate mg	Biotin µg	Vitamin C mg
193	**Chicken in sauce with vegetables,** chilled/frozen, *reheated*	12	320	N	N	0.26	0.16	9.1	N	0.27	Tr	12	1.08	3	Tr
194	**Chicken in white sauce,** canned	Tr	Tr	N	N	0.01	0.10	2.3	N	0.10	Tr	5	N	N	0
195	-, made with whole milk	53	28	0.4	0.79	0.06	0.17	5.2	3.4	0.23	Tr	7	0.91	3	0
196	-, made with semi-skimmed milk	38	23	0.4	0.76	0.06	0.18	5.2	3.4	0.23	Tr	7	0.90	3	0
197	**Chicken korma**	18	(14)	0.1	(0.13)	0.10	0.12	5.0	3.0	0.28	Tr	7	0.69	2	1
198	**Chicken risotto**	21	26	0.1	N	0.09	0.05	3.0	2.0	N	Tr	N	N	N	1
199	**Chicken, stir-fried with mushrooms and cashew nuts**	Tr	845	0.1	(0.27)	0.13	0.12	6.9	3.9	(0.40)	Tr	12	N	N	15
200	**Chicken, stir-fried with peppers in black bean sauce**	Tr	83	0.1	(0.26)	0.18	0.09	6.0	3.4	0.36	Tr	10	0.72	N	16
201	**Chicken, stir-fried with rice and vegetables,** frozen, *reheated*	Tr	565	N	N	0.09	0.09	1.9	1.3	0.22	Tr	21	0.50	4	2
202	**Chicken vindaloo**	8	36	0.1	(0.13)	0.12	0.12	5.0	3.5	0.27	Tr	6	0.72	2	1
203	-, reduced fat	9	29	0.1	(0.14)	0.11	0.12	5.3	3.6	0.29	Tr	6	0.78	2	1
204	**Chicken wings,** marinated, chilled/frozen, *barbecued*	(24)	N	(0.1)	(0.23)	(0.07)	(0.11)	(6.2)	(5.3)	(0.27)	(1)	(10)	(1.34)	(4)	Tr
205	-, *weighed with bone*	(16)	N	(0.1)	(0.15)	(0.05)	(0.07)	(4.0)	(3.5)	(0.18)	(1)	(7)	(0.87)	(3)	Tr
206	**Chilli con carne**	Tr	115	0.2	(0.56)	0.07	0.06	2.2	1.6	0.19	1	8	0.25	N	7

Meat dishes *continued*

No. 19-	Food	Description and main data sources	Water g	Total Nitrogen g	Protein g	Fat g	Carbo-hydrate g	Energy value kcal	kJ
207	**Chilli con carne**, canned	Manufacturers' data, 10 brands	N	1.31	8.2	4.7	11.7	119	501
208	-, chilled/frozen, *reheated*	11 samples	76.4	1.23	7.7	4.3	7.1	96	404
209	-, -, with rice	Calculated from 60% chilli con carne (No 208) and 40% boiled white rice	73.8	0.88	5.5	2.7	16.1	107	451
210	**Coq au vin**	Recipe	(68.6)	1.78	11.1	11.0	3.2	155	647
211	-, *weighed with bone*	Recipe[a]	56.0	1.44	9.0	9.0	2.6	127	528
212	**Corned beef hash**	Recipe	69.4	1.68	10.4	5.9	12.3[b]	141	592
213	**Coronation chicken**	Recipe	46.7	2.66	16.6	31.7	3.2	364	1506
214	-, reduced fat	Recipe	62.7	2.66	16.6	14.8	4.2	215	897
215	**Cottage pie**	Recipe	74.4	1.02	6.4	6.8	10.4[c]	126	527
216	**Cottage/Shepherd's Pie**, chilled/frozen, *reheated*	11 samples including beef and lamb. 11.5-25% meat	73.1	0.72	4.5	5.4	11.9	111	467
217	**Devilled kidneys**	Recipe	73.1	2.18	13.6	8.2	2.0	136	567
218	**Duck à l'orange**, excluding fat and skin	Recipe	75.1	1.87	11.7	6.8	3.9	123	513
219	-, including fat and skin	Recipe	64.3	1.63	10.2	20.9	3.7	243	1006
220	**Duck with pineapple**	Recipe	53.2	2.22	13.9	25.7	6.2[b]	310	1286
221	**Goulash**	Recipe	80.1	1.10	6.9	3.4	6.6[b]	83	349
222	**Irish stew**	Recipe	76.2	1.22	7.6	6.4	8.9[d]	121	508
223	-, made with lean beef	Recipe	77.5	1.28	8.0	5.1	8.9[d]	111	467
224	-, canned	10 samples, 2 brands	82.5	0.75	4.7	5.1	6.8	91	379

[a] Edible proportion of the whole dish is 0.82
[c] Includes 0.1g oligosaccharides per 100g food
[b] Includes 0.4g oligosaccharides per 100g food
[d] Includes 0.3g oligosaccharides per 100g food

Meat dishes continued

No. 19-	Food	Starch g	Total sugars g	Dietary fibre Southgate method g	Dietary fibre Englyst method g	Satd g	Fatty acids cis & trans Mono- unsatd g	Poly- unsatd g	Total trans g	Cholesterol mg
207	**Chilli con carne**, canned	(9.2)	2.5	N	(2.4)	1.6	N	N	N	N
208	-, chilled/frozen, *reheated*	4.4	2.7	N	1.4	1.9	1.9	0.2	0.2	N
209	-, -, with rice	14.5	1.6	N	0.9	1.1	1.1	0.1	0.1	N
210	**Coq au vin**	2.8	0.4	0.4	0.3	4.3	4.4	1.6	0.2	69
211	-, *weighed with bone*	2.3	0.3	0.3	0.2	3.5	3.6	1.3	0.2	56
212	**Corned beef hash**	10.1	1.8	1.1	1.0	3.3	2.0	0.3	0.3	35
213	**Coronation chicken**	0.2	3.0	N	N	5.2	8.2	16.5	0.1	89
214	-, reduced fat	1.4	2.8	N	Tr	N	N	N	0.1	70
215	**Cottage pie**	8.7	1.6	1.0	0.9	2.5	2.7	0.9	0.4	19
216	**Cottage/Shepherd's pie**, chilled/frozen, *reheated*	10.3	1.6	N	0.9	2.4	2.2	0.4	0.3	16
217	**Devilled kidneys**	0.1	1.9	0.4	0.3	3.2	2.4	1.7	Tr	240
218	**Duck à l'orange**, excluding fat and skin	3.0	0.9	Tr	Tr	2.7	2.8	0.7	0.2	58
219	-, including fat and skin	2.8	0.9	Tr	Tr	6.8	10.0	2.7	0.3	54
220	**Duck with pineapple**	1.3	4.6	0.3	0.3	7.7	13.0	3.6	0.3	66
221	**Goulash**	4.4	1.8	0.9	0.8	0.8	1.2	0.8	0.1	18
222	**Irish stew**	6.7	1.9	1.1	1.0	2.9	2.4	0.4	0.5	28
223	-, made with lean beef	6.7	1.9	1.1	1.0	2.2	1.9	0.3	0.4	26
224	-, canned	5.6	1.2	0.1	N	2.5	2.0	0.3	0.4	15

Meat dishes *continued*

No. 19-	Food	Na	K	Ca	Mg	P	Fe	Cu	Zn	Cl	Mn	Se	I
							mg					µg	
207	**Chilli con carne**, canned	690	N	N	N	N	N	N	N	N	N	N	N
208	-, chilled/frozen, *reheated*	310	300	42	21	90	1.5	0.13	1.4	(470)	0.18	N	N
209	-, -, with rice	180	190	26	14	68	1.0	0.10	1.0	(290)	0.23	2	N
210	**Coq au vin**	260	280	16	19	87	1.1	0.11	0.9	390	0.09	8	N
211	-, *weighed with bone*	210	230	13	15	70	0.8	0.10	0.8	320	0.08	7	N
212	**Corned beef hash**	400	250	20	15	72	1.1	0.12	2.2	720	0.09	(4)	9
213	**Coronation chicken**	240	210	12	17	140	0.8	0.07	0.9	380	0.03	N	17
214	-, reduced fat	410	N	N	N	N	N	N	N	630	N	N	N
215	**Cottage pie**	290	240	17	13	66	0.7	0.04	1.2	320	0.08	(3)	5
216	**Cottage/Shepherd's pie,** chilled/frozen, *reheated*	420	240	20	14	65	0.7	0.04	0.9	710	0.08	N	N
217	**Devilled kidneys**	370	330	31	21	230	4.5	0.68	2.0	580	0.18	110	N
218	**Duck à l'orange,** excluding fat and skin	360	150	1	11	97	1.4	0.15	1.2	320	Tr	(1)	N
219	-, including fat and skin	330	130	15	10	96	1.0	0.12	1.1	290	0.11	(11)	N
220	**Duck with pineapple**	300	200	22	16	130	1.4	0.18	1.5	340	0.30	(15)	N
221	**Goulash**	380	280	13	15	71	1.0	0.06	1.5	350	0.08	2	5
222	**Irish stew**	200	280	14	14	82	0.8	0.06	1.5	220	0.08	1	4
223	-, made with lean beef	200	290	13	15	86	0.8	0.06	1.6	220	0.08	1	4
224	-, canned	280	130	10	8	40	1.2	0.13	0.7	(430)	0.05	N	N

Meat dishes *continued*

19-207 to 19-224
Vitamins per 100g food

No. 19-	Food	Retinol µg	Carotene µg	Vitamin D µg	Vitamin E mg	Thiamin mg	Ribo-flavin mg	Niacin mg	Trypt 60 mg	Vitamin B6 mg	Vitamin B12 µg	Folate µg	Panto-thenate mg	Biotin µg	Vitamin C mg
207	**Chilli con carne**, canned	N	N	N	N	N	N	N	N	N	N	N	N	N	N
208	-, chilled/frozen, *reheated*	65	99	N	N	0.07	0.12	1.6	1.3	0.19	1	15	0.51	4	N
209	-, -, with rice	39	59	N	N	0.05	0.08	1.1	1.0	0.13	Tr	10	0.39	3	N
210	**Coq au vin**	45	21	0.3	N	0.10	0.15	2.6	2.0	0.16	Tr	5	0.60	3	1
211	-, *weighed with bone*	37	17	0.3	N	0.08	0.12	2.1	1.6	0.13	Tr	4	0.49	2	Tr
212	**Corned beef hash**	20	12	0.5	0.33	0.11	0.06	1.1	2.6	0.24	Tr	10	0.32	1	2
213	**Coronation chicken**	37	36	0.2	6.88	0.05	0.12	5.5	3.3	0.22	Tr	7	N	N	1
214	-, reduced fat	7	21	0.1	3.10	N	N	N	N	N	Tr	N	N	N	1
215	**Cottage pie**	15	790	0.3	0.39	0.09	0.03	1.2	1.2	0.21	Tr	8	0.26	1	2
216	**Cottage/Shepherd's pie,** chilled/frozen, *reheated*	17	110	0.1[a]	0.28	0.15	0.10	1.3	0.8	0.19	1	14	0.39	1	1
217	**Devilled kidneys**	130	41	0.4	(0.41)	0.23	1.48	6.0	2.8	0.38	10	8	2.90	27	4
218	**Duck à l'orange,** excluding fat and skin	N	11	N	0.07	0.12	0.21	2.3	2.4	0.12	1	5	0.69	2	3
219	-, including fat and skin	N	N	N	N	0.09	0.25	1.9	2.1	0.16	1	8	1.30	3	3
220	**Duck with pineapple**	N	N	N	N	0.12	0.28	2.2	2.9	0.20	1	7	1.42	4	2
221	**Goulash**	Tr	77	0.2	(0.34)	0.08	(0.07)	1.2	1.3	0.24	Tr	9	0.26	1	8
222	**Irish stew**	Tr	1200	0.2	0.13	0.12	0.05	1.4	1.5	0.22	Tr	8	0.39	1	8
223	-, made with lean beef	Tr	1200	0.1	0.12	0.13	0.05	1.4	1.6	0.22	1	8	0.40	1	3
224	-, canned	Tr	N	N	N	0.02	0.06	0.9	N	0.14	Tr	3	N	N	N

[a] Contribution from 25-hydroxycholecalciferol not included

83

No. 19-	Food	Description and main data sources	Water g	Total Nitrogen g	Protein g	Fat g	Carbohydrate g	Energy value kcal	Energy value kJ
225	**Lamb biryani**	Recipe	60.8	1.15	7.2	10.5	20.5[a]	200	839
226	**-, reduced fat**	Recipe	65.3	1.24	7.6	4.5	21.4[a]	151	638
227	**Lamb curry,** made with canned curry sauce	Recipe	59.3	2.48	15.5	19.2	4.4	251	1044
228	**Lamb kheema**	Recipe	68.8	1.71	10.7	14.5	3.2[b]	185	770
229	**-, reduced fat**	Recipe	72.8	1.78	11.1	9.7	3.3[b]	144	600
230	**Lamb koftas**	Recipe	51.6	3.76	23.5	16.1	6.3[c]	263	1096
231	**Lamb/Beef hot pot with potatoes,** chilled/frozen, retail, *reheated*	10 samples, 6 brands of beef and Lancashire hot pot. 10-32% meat	74.4	1.15	7.2	4.4	10.6	108	455
232	**Lamb rogan josh**	Recipe	68.1	2.30	14.4	9.5	4.0[d]	158	660
233	**Lamb, stir-fried with vegetables**	Recipe	62.4	2.59	16.2	18.4	2.0	238	988
234	**Lamb vindaloo**	Recipe	46.6	3.17	19.8	13.5	2.6[e]	210	876
235	**Lamb's heart casserole**	Recipe	67.5	1.94	12.1	7.8	10.0[f]	156	654
236	**Lancashire hot pot**	Recipe	76.5	1.42	8.9	5.2	7.5[g]	111	464
237	**Lasagne**	Recipe	71.3	1.36	8.5	10.8	7.4[f]	159	663
238	**-, chilled/frozen,** *reheated*	12 samples, 11 brands. 10-20% meat	68.1	1.18	7.4	6.1	15.7	143	603
239	**Lemon chicken**	Recipe	67.4	2.62	16.4	6.1	9.2	155	652
240	**Liver and bacon,** *fried*	Recipe	52.8	4.60	28.8	14.6	Tr	247	1030
241	**Liver and onions,** *stewed*	Recipe	67.7	2.37	14.8	7.6	5.4[h]	148	619

[a] Includes 0.5g oligosaccharides per 100g food
[b] Includes 0.6g oligosaccharides per 100g food
[c] Includes 0.3g oligosaccharides per 100g food
[d] Includes 0.7g oligosaccharides per 100g food
[e] Includes 0.4g oligosaccharides per 100g food
[f] Includes 0.1g oligosaccharides per 100g food
[g] Includes 0.2g oligosaccharides per 100g food
[h] Includes 1.6g oligosaccharides per 100g food

Meat dishes continued

| No. 19- | Food | Starch g | Total sugars g | Dietary fibre | | Fatty acids cis & trans | | | Total trans g | Cholesterol mg |
				Southgate method g	Englyst method g	Satd g	Mono-unsatd g	Poly-unsatd g		
225	**Lamb biryani**	18.1	1.9	N	0.6	2.1	3.7	3.8	0.2	16
226	-, reduced fat	19.0	1.9	N	0.7	1.4	1.6	0.9	0.2	17
227	**Lamb curry**, made with canned curry sauce	2.1	2.3	N	N	N	N	N	1.2	63
228	**Lamb kheema**	1.1	1.5	1.6	(1.2)	3.8	5.3	4.2	0.5	36
229	-, reduced fat	1.1	1.6	(1.7)	(1.3)	3.4	3.6	1.8	0.5	38
230	**Lamb koftas**	5.1	0.9	0.6	0.3	7.2	6.3	0.9	1.2	125
231	**Lamb/Beef hot pot with potatoes**, chilled/frozen, retail, reheated	9.6	1.0	N	0.9	1.7	1.9	0.5	0.4	N
232	**Lamb rogan josh**	0.3	3.0	0.8	0.8	3.7	3.6	1.3	0.6	45
233	**Lamb, stir-fried with vegetables**	0.3	1.7	N	0.6	7.4	7.0	2.5	1.3	66
234	**Lamb vindaloo**	0.4	1.8	(0.2)	(0.3)	5.2	5.3	1.9	0.8	3
235	**Lamb's heart casserole**	9.1	0.8	0.7	0.4	3.0	2.4	1.1	0.1	93
236	**Lancashire hot pot**	5.9	1.4	1.0	0.9	2.1	1.9	0.6	0.4	27
237	**Lasagne**	4.8	2.5	0.7	0.5	4.7	3.9	1.3	0.4	29
238	-, chilled/frozen, reheated	12.7	3.0	N	0.7	2.8	2.2	0.7	0.3	18
239	**Lemon chicken**	4.9	4.3	Tr	Tr	0.8	2.2	2.7	Tr	46
240	**Liver and bacon**, fried	0	0	0	0	N	N	N	N	331
241	**Liver and onions**, stewed	Tr	3.8	1.0	1.0	1.5	2.4	2.2	Tr	295

Meat dishes continued

Inorganic constituents per 100g food

No. 19-	Food	Na	K	Ca	Mg	P	Fe	Cu	Zn	Cl	Mn	Se	I
						mg						µg	
225	**Lamb biryani**	N	180	33	14	82	1.0	0.10	1.1	N	0.29	5	9
226	**-**, reduced fat	N	190	34	15	85	1.1	0.10	1.1	N	0.31	5	N
227	**Lamb curry**, made with canned curry sauce	650	350	22	26	160	1.6	0.09	3.1	520	0.13	N	N
228	**Lamb kheema**	230	270	26	22	120	1.5	0.08	1.9	340	0.13	(1)	5
229	**-**, reduced fat	240	280	27	22	120	1.5	0.08	2.0	360	0.14	1	5
230	**Lamb koftas**	360	410	47	31	240	2.5	0.14	4.1	510	0.11	7	14
231	**Lamb/Beef hot pot with potatoes**, chilled/frozen, retail, *reheated*	330	260	17	16	80	0.8	0.07	1.3	(510)	0.08	N	N
232	**Lamb rogan josh**	170	380	40	24	120	1.9	0.11	2.0	270	0.14	(1)	(5)
233	**Lamb, stir-fried with vegetables**	370	370	20	22	160	1.8	0.08	3.3	550	0.10	N	N
234	**Lamb vindaloo**	220	330	60	31	170	3.0	0.12	2.8	310	0.22	(2)	(4)
235	**Lamb's heart casserole**	430	210	26	18	150	2.5	0.35	1.3	500	0.12	(2)	N
236	**Lancashire hot pot**	200	260	18	13	77	0.7	0.06	1.2	210	0.06	(1)	N
237	**Lasagne**	350	190	97	14	120	0.7	0.06	1.3	390	0.09	4	10
238	**-**, chilled/frozen, *reheated*	390	230	80	19	120	1.0	0.10	1.4	(600)	0.22	N	N
239	**Lemon chicken**	180	270	5	21	150	0.4	0.05	0.5	260	0.01	N	5
240	**Liver and bacon**, *fried*	640	340	8	25	430	8.8	10.8	5.1	1100	0.36	53	2
241	**Liver and onions**, *stewed*	280	310	22	16	290	5.4	6.67	2.9	440	0.29	(29)	6

86

No. 19-	Food	Retinol µg	Carotene µg	Vitamin D µg	Vitamin E mg	Thiamin mg	Ribo-flavin mg	Niacin mg	Trypt 60 mg	Vitamin B6 mg	Vitamin B12 µg	Folate µg	Panto-thenate mg	Biotin µg	Vitamin C mg
225	**Lamb biryani**	4	28	0.1	N	0.06	0.05	1.1	1.2	0.14	Tr	6	0.21	1	1
226	-, reduced fat	3	28	0.1	0.16	0.07	0.05	1.1	1.2	0.15	Tr	6	N	1	1
227	**Lamb curry**, made with canned curry sauce	6	N	0.3	N	0.09	0.12	2.5	2.9	0.14	1	N	N	N	Tr
228	**Lamb kheema**	Tr	(130)	0.3	(0.32)	0.11	0.09	2.3	2.0	0.12	1	10	0.40	1	4
229	-, reduced fat	Tr	140	0.3	(0.33)	0.12	0.09	2.4	2.1	0.13	1	10	0.42	1	4
230	**Lamb koftas**	26	18	0.8	(0.28)	0.15	0.21	4.5	4.7	0.21	2	7	0.98	4	2
231	**Lamb/Beef hot pot with potatoes**, chilled/frozen, retail, *reheated*	N	N	N	N	0.43	0.09	1.5	N	0.29	1	25	N	N	Tr
232	**Lamb rogan josh**	6	160	0.3	(0.56)	0.10	0.10	2.4	2.1	0.20	1	6	(0.29)	(1)	4
233	**Lamb, stir-fried with vegetables**	7	230	0.4	N	0.11	0.12	2.8	3.1	0.18	1	11	0.59	N	5
234	**Lamb vindaloo**	8	39	0.4	(0.09)	0.09	0.11	2.8	2.8	(0.19)	1	3	(0.28)	N	1
235	**Lamb's heart casserole**	23	14	0.1	(0.24)	0.26	0.42	3.5	2.5	0.16	4	4	1.25	3	2
236	**Lancashire hot pot**	Tr	565	0.2	(0.08)	0.10	0.05	1.3	1.3	0.21	1	8	0.25	1	3
237	**Lasagne**	52	355	0.3	(0.58)	0.06	0.10	1.4	1.7	0.12	1	7	0.31	2	1
238	-, chilled/frozen, *reheated*	Tr	N	N	N	0.33	0.12	1.4	1.3	0.14	1	11	0.38	4	N
239	**Lemon chicken**	Tr	Tr	0.1	(0.07)	0.08	0.09	5.7	3.2	0.27	Tr	5	0.69	1	1
240	**Liver and bacon**, *fried*	15700	48	0.5	0.22	0.38	3.62	13.7	4.8	0.42	53	104	5.29	25	7
241	**Liver and onions**, *stewed*	11900	65	0.3	(0.59)	0.29	2.55	9.4	3.2	0.37	30	75	4.56	26	8

No. 19-	Food	Description and main data sources	Water g	Total Nitrogen g	Protein g	Fat g	Carbohydrate g	Energy value kcal	kJ
242	**Minced beef**, *stewed*	Recipe	73.6	1.78	11.1	10.4	2.7[a]	148	617
243	-, extra lean, *stewed*	Recipe	76.9	1.95	12.2	6.9	2.6[a]	121	504
244	**Minced beef with gravy** with/without onions, canned	Manufacturers' data, 11 brands	N	1.66	10.4	11.8	5.6	169	703
245	**Minced beef with vegetables,** *stewed*	Recipe	75.1	1.50	9.4	7.3	5.8[b]	125	523
246	**Minced lamb,** *stewed*	Recipe	76.3	1.71	10.7	8.9	2.7[a]	133	554
247	**Moussaka**	Recipe	76.6	1.36	8.5	7.9	4.5[a]	122	509
248	-, chilled/frozen/longlife, *reheated*	8 samples, 4 brands of beef and lamb. 20-23% meat	70.6	1.33	8.3	8.3	8.6[c]	140	586
249	**Pancakes, beef,** frozen, *shallow-fried*	Coated crispy pancakes. 4 samples, 2 brands. 15-16% meat	50.1	1.14	7.1	15.7	24.7	262	1097
250	-, **chicken**, frozen, *shallow-fried*	Coated crispy pancakes. 4 samples of the same brand. 10% meat	49.9	0.99	6.2	14.2	28.7	260	1090
251	**Pasta with ham and mushroom sauce**	Recipe	74.6	0.91	5.7	6.0	11.7	121	508
252	**Pasta with meat and tomato sauce**	Recipe	74.5	1.06	6.6	4.2	12.6	112	469
253	**Pork and apple casserole**	Recipe	76.6	1.74	10.9	3.1	7.3[a]	99	417

[a] Includes 0.2g oligosaccharides per 100g food
[c] Includes 0.1g oligosaccharides per 100g food
[b] Includes 0.4g oligosaccharides per 100g food

Meat dishes continued

Composition of food per 100g

No. 19-	Food	Starch g	Total sugars g	Dietary fibre Southgate method g	Dietary fibre Englyst method g	Fatty acids Satd g	Fatty acids cis & trans Mono-unsatd g	Fatty acids cis & trans Poly-unsatd g	Total trans g	Cholesterol mg
242	**Minced beef**, stewed	1.8	0.7	0.2	0.2	3.9	4.3	1.0	0.4	32
243	-, extra lean, stewed	1.7	0.7	0.2	0.2	2.4	2.7	0.9	0.2	30
244	**Minced beef with gravy** with/without onions, canned	(4.6)	1.0	N	(0.3)	5.3	N	N	N	N
245	**Minced beef with vegetables**, stewed	3.1	2.3	0.8	0.8	3.1	3.1	0.3	0.3	26
246	**Minced lamb**, stewed	1.8	0.7	0.2	0.2	3.5	3.3	1.1	0.6	41
247	**Moussaka**	1.5	2.8	1.1	1.0	3.6	2.8	0.8		44
248	-, chilled/frozen/longlife, reheated	6.5	2.0	N	0.8	2.9	3.6	1.1	0.4	26
249	**Pancakes, beef**, frozen, shallow-fried	22.2	2.5	N	1.0	2.2	6.5	4.9	0.1	23
250	-, **chicken**, frozen, shallow-fried	25.2	3.5	N	0.1	1.4	5.6	4.5	0.1	14
251	**Pasta with ham and mushroom sauce**	10.9	0.8	1.5	0.9	(3.5)	(1.6)	(0.4)	Tr	20
252	**Pasta with meat and tomato sauce**	10.7	1.9	N	N	1.7	1.6	0.5	0.2	13
253	**Pork and apple casserole**	2.9	4.2	0.5	0.5	0.7	1.1	0.9	Tr	28

No. 19-	Food	Na	K	Ca	Mg	P	Fe (mg)	Cu	Zn	Cl	Mn	Se (µg)	I
242	**Minced beef**, *stewed*	360	160	13	11	94	0.9	0.01	2.1	270	0.05	(4)	6
243	-, extra lean, *stewed*	370	180	13	12	100	1.0	0.04	2.4	270	0.03	4	7
244	**Minced beef with gravy** with/without onions, canned	400	N	N	N	N	N	N	N	N	N	N	N
245	**Minced beef with vegetables**, *stewed*	360	180	20	1	84	0.9	0.02	1.8	300	0.08	4	5
246	**Minced lamb**, *stewed*	350	190	17	13	110	1.0	0.05	1.9	260	0.04	1	4
247	**Moussaka**	200	270	63	17	110	0.9	0.06	1.4	310	0.09	(2)	(9)
248	-, chilled/frozen/longlife, *reheated*	350	250	75	77	110	0.6	0.12	0.1	(540)	0.13	N	N
249	**Pancakes**, **beef**, frozen, *shallow-fried*	640	190	70	16	95	1.1	0.06	1.0	970	0.21	N	N
250	-, **chicken**, frozen, *shallow-fried*	680	170	85	16	100	0.9	0.05	0.5	1080	0.22	N	N
251	**Pasta with ham and mushroom sauce**	150	140	75	14	110	0.6	0.27	0.8	240	0.16	(3)	N
252	**Pasta with meat and tomato sauce**	140	220	12	18	71	0.8	0.10	1.2	270	0.19	N	N
253	**Pork and apple casserole**	260	240	15	15	110	0.6	0.03	0.9	240	0.07	7	3

Meat dishes *continued*

No. 19-	Food	Retinol µg	Carotene µg	Vitamin D µg	Vitamin E mg	Thiamin mg	Ribo-flavin mg	Niacin mg	Trypt 60 mg	Vitamin B6 mg	Vitamin B12 µg	Folate µg	Panto-thenate mg	Biotin µg	Vitamin C mg
242	**Minced beef**, *stewed*	Tr	Tr	0.2	0.10	0.04	0.06	2.6	2.0	0.18	1	5	0.22	1	Tr
243	-, extra lean, *stewed*	Tr	Tr	0.3	(0.11)	0.05	0.06	2.8	2.2	0.20	1	5	0.25	1	Tr
244	**Minced beef with gravy** with/without onions, canned	N	N	N	N	N	N	N	N	N	N	N	N	N	N
245	**Minced beef with vegetables**, *stewed*	Tr	1400	0.2	0.19	0.06	0.05	2.2	1.7	0.18	1	6	0.23	1	1
246	**Minced lamb**, *stewed*	Tr	Tr	0.3	(0.11)	0.07	0.08	2.1	2.0	0.10	(1)	2	0.40	1	Tr
247	**Moussaka**	40	100	0.4	0.65	0.07	0.10	1.4	1.7	0.12	1	7	0.40	N	2
248	-, chilled/frozen/longlife, *reheated*	40	235	0.3	N	0.05	0.19	1.5	1.5	0.15	1	8	0.48	2	N
249	**Pancakes, beef**, frozen, *shallow-fried*	Tr	Tr	N	1.72	0.20	0.11	1.1	1.3	0.14	1	20	0.58	3	Tr
250	-, **chicken**, frozen, *shallow-fried*	Tr	Tr	N	2.73	0.15	0.12	1.3	1.0	0.12	Tr	8	0.61	2	Tr
251	**Pasta with ham and mushroom sauce**	69	32	0	0.15	0.06	0.15	1.4	1.2	0.07	Tr	9	0.50	4	Tr
252	**Pasta with meat and tomato sauce**	Tr	(30)	0.1	N	0.04	0.14	1.3	1.2	0.08	Tr	5	N	N	Tr
253	**Pork and apple casserole**	Tr	6	0.2	N	0.36	0.09	2.8	2.2	0.24	Tr	2	0.58	1	2

Composition of food per 100g

No. 19-	Food	Description and main data sources	Water g	Total Nitrogen g	Protein g	Fat g	Carbo-hydrate g	Energy value kcal	kJ
254	**Pork and beef meatballs in tomato sauce**	Recipe	75.1	1.62	10.1	7.6	4.5[a]	126	525
255	**Pork and chicken chow mein**	Recipe	80.2	1.26	7.9	2.9	6.1	81	339
256	**Pork casserole**, made with canned cook-in sauce	Recipe	70.1	2.74	17.1	7.8	3.8	153	640
257	**Pork chops in mustard and cream**	Recipe	(57.0)	2.32	14.5	21.6	2.4[a]	261	1084
258	-, *weighed with bone*	Recipe[b]	(49.9)	2.06	12.9	19.0	2.0[c]	230	954
259	**Pork and pineapple kebabs**	Recipe	64.3	2.42	15.1	9.0	7.7	170	713
260	**Pork spare ribs, 'barbecue style'**	Recipe	(38.5)	3.30	20.6	24.2	4.8[d]	318	1322
261	-, *weighed with bone*	Recipe[e]	(18.1)	1.55	9.7	11.3	2.3[f]	149	620
262	-, *chilled/frozen, reheated*	5 samples including American and Chinese style and hot and spicy ribs. 80-90% meat	49.6	4.21	26.3	17.1	5.8	281	117
263	-, -, *weighed with bone*	Calculated from No 262[g]	17.4	1.47	9.2	6.0	2.0	98	410
264	**Pork spare ribs in black bean sauce**	Recipe	65.3	2.40	15.0	10.1	5.4	171	715
265	-, *weighed with bone*	Recipe[h]	68.3	2.64	16.5	10.7	3.0	174	724
266	**Pork, stir-fried with vegetables**	Recipe	77.1	1.94	12.1	4.7	3.9	105	442
267	**Rabbit casserole**	Recipe	78.7	1.86	11.6	5.1	2.6	102	428
268	**Salmis of pheasant**	Recipe	78.0	1.71	10.7	6.1	2.6[c]	108	449

[a] Includes 0.2g oligosaccharides per 100g food
[d] Includes 0.8g oligosaccharides per 100g food
[g] Edible proportion is 0.35
[b] Edible proportion of the whole dish is 0.84
[e] Edible proportion of the whole dish is 0.47
[h] Edible proportion of the whole dish is 0.57
[c] Includes 0.1g oligosaccharides per 100g food
[f] Includes 0.4g oligosaccharides per 100g food

Meat dishes *continued*

Composition of food per 100g

No. 19-	Food	Starch g	Total sugars g	Dietary fibre Southgate method g	Dietary fibre Englyst method g	Fatty acids Satd g	cis & trans Mono-unsatd g	Poly-unsatd g	Total trans g	Cholesterol mg
254	**Pork and beef meatballs in tomato sauce**	1.6	2.7	(1.0)	(0.8)	2.8	3.0	1.1	0.2	43
255	**Pork and chicken chow mein**	4.2	1.9	(1.3)	(0.6)	0.7	1.1	0.8	Tr	29
256	**Pork casserole**, made with canned cook-in sauce	1.5	2.3	0	0	2.6	3.2	1.4	0.1	50
257	**Pork chops in mustard and cream**	0.1	2.1	0.1	0.3	8.3	8.1	3.6	0.1	55
258	-, *weighed with bone*	Tr	1.9	0.1	0.2	7.3	7.1	3.2	0.1	49
259	**Pork and pineapple kebabs**	0.1	7.6	1.2	0.9	2.1	3.2	3.2	Tr	41
260	**Pork spare ribs, 'barbecue style'**	0.6	3.4	0.5	0.6	6.4	9.2	6.6	0.1	69
261	-, *weighed with bone*	0.3	1.6	0.3	0.3	3.0	4.2	3.1	Tr	32
262	-, *chilled/frozen, reheated*	1.4a	4.4	Tr	Tr	6.2	6.8	2.7	0.1	(160)
263	-, -, *weighed with bone*	0.5	1.5	Tr	Tr	2.2	2.4	1.0	Tr	(57)
264	**Pork spare ribs in black bean sauce**	0.7	4.7	N	0.3	3.8	3.9	1.2	0.1	48
265	-, *weighed with bone*	0.4	2.6	N	0.2	3.9	4.0	1.9	0.1	57
266	**Pork, stir-fried with vegetables**	2.3	1.6	(2.6)	0.9	1.3	1.6	1.4	Tr	30
267	**Rabbit casserole**	1.5	1.1	0.4	0.4	1.5	1.4	1.6	0.1	28
268	**Salmis of pheasant**	0.9	1.6	0.4	0.4	2.0	2.6	1.0	0.2	83

a Includes maltodextrins

Meat dishes continued

No. 19-	Food	Na	K	Ca	Mg	P	Fe	Cu	Zn	Cl	Mn	Se	I
		mg										µg	
254	Pork and beef meatballs in tomato sauce	240	290	46	18	110	1.1	0.06	1.5	370	(0.11)	N	(6)
255	Pork and chicken chow mein	440	200	13	15	84	0.8	0.10	0.7	(600)	(0.10)	N	N
256	Pork casserole, made with canned cook-in sauce	480	340	8	20	170	0.7	0.03	1.4	320	0.03	10	4
257	Pork chops in mustard and cream	310	280	35	22	150	0.8	0.06	1.1	370	0.07	8	N
258	-, weighed with bone	270	250	31	20	130	0.7	0.05	1.0	320	0.07	7	N
259	Pork and pineapple kebabs	210	400	11	24	170	0.9	0.18	1.2	320	0.28	11	4
260	Pork spare ribs, 'barbecue style'	1160	420	35	28	180	1.7	0.07	2.8	1680	0.11	N	N
261	-, weighed with bone	540	200	16	13	85	0.8	0.04	1.3	790	0.05	N	N
262	-, chilled/frozen, reheated	440	380	70	27	220	1.6	0.15	3.1	640	0.16	(25)	(4)
263	-, -, weighed with bone	150	130	25	9	75	0.6	0.05	1.1	220	0.06	(8)	(1)
264	Pork spare ribs in black bean sauce	930	270	25	24	130	1.6	0.06	2.0	1390	0.08	9	4
265	-, weighed with bone	530	320	15	24	170	1.2	0.08	2.2	790	0.05	10	4
266	Pork, stir-fried with vegetables	330	260	16	24	140	1.3	0.10	1.0	250	0.16	N	N
267	Rabbit casserole	490	220	17	14	120	0.7	0.04	0.8	450	0.03	9	N
268	Salmis of pheasant	230	180	23	13	90	1.1	0.05	0.5	190	0.04	5	1

Meat dishes continued

19-254 to 19-268
Vitamins per 100g food

No. 19-	Food	Retinol μg	Carotene μg	Vitamin D μg	Vitamin E mg	Thiamin mg	Ribo-flavin mg	Niacin mg	Trypt 60 mg	Vitamin B6 mg	Vitamin B12 μg	Folate μg	Panto-thenate mg	Biotin μg	Vitamin C mg
254	Pork and beef meatballs in tomato sauce	15	645	0.3	(0.59)	0.19	0.09	(2.4)	1.9	0.21	1	(9)	0.43	N	(20)
255	Pork and chicken chow mein	Tr	460	0.1	(0.16)	0.13	0.08	2.0	(1.6)	(0.14)	Tr	(7)	N	N	4
256	Pork casserole, made with canned cook-in sauce	Tr	Tr	0.5	0.03	0.54	0.14	4.3	3.4	0.35	1	1	0.90	1	0
257	Pork chops in mustard and cream	61	20	0.7	N	0.48	0.13	2.9	2.6	0.38	1	2	0.62	3	Tr
258	-, weighed with bone	54	18	0.6	N	0.42	0.11	2.6	2.3	0.33	1	2	0.54	2	Tr
259	Pork and pineapple kebabs	Tr	170	0.3	N	0.64	0.21	3.8	3.0	0.37	Tr	10	N	N	17
260	Pork spare ribs, 'barbecue style'	Tr	4	0.8	(0.10)	0.68	0.16	4.7	3.8	0.34	1	5	1.31	Tr	1
261	-, weighed with bone	Tr	Tr	0.4	(0.04)	0.31	0.07	2.2	1.8	0.16	Tr	2	0.60	Tr	1
262	-, chilled/frozen, reheated	7	220	(1.7)	0.05	0.82	0.38	7.0	(7.3)	0.30	(1)	7	(2.64)	(4)	Tr
263	-, -, weighed with bone	Tr	77	(0.6)	0.02	0.29	0.13	2.5	(2.6)	0.10	Tr	3	(0.92)	(1)	Tr
264	Pork spare ribs in black bean sauce	Tr	Tr	0.5	0.01	0.56	0.13	3.2	2.7	0.20	1	1	0.90	1	Tr
265	-, weighed with bone	Tr	Tr	0.5	0.01	0.58	0.20	3.6	3.3	0.28	1	4	0.86	2	Tr
266	Pork, stir-fried with vegetables	Tr	78	0.3	N	0.48	0.14	2.5	2.4	0.26	Tr	16	0.70	N	8
267	Rabbit casserole	N	N	N	(0.12)	0.07	0.08	3.4	2.1	0.22	4	2	0.37	1	1
268	Salmis of pheasant	12	315	0.1	0.32	0.03	0.09	2.8	2.2	0.19	1	6	0.34	Tr	3

Meat dishes *continued*

Composition of food per 100g

No. 19-	Food	Description and main data sources	Water g	Total Nitrogen g	Protein g	Fat g	Carbohydrate g	Energy value kcal	kJ
269	**Sausage casserole**	Recipe	68.5	1.90	11.9	10.9	5.1[a]	165	687
270	**Shepherd's pie**	Recipe	77.2	0.96	6.0	5.9	9.3	112	469
271	**Shish kebabs, with onions and peppers**	Recipe	62.4	2.06	12.9	16.2	3.9[b]	212	881
272	**Spaghetti bolognese**	Recipe	(72.1)	1.25	7.8	5.6	12.5[c]	129	540
273	-, chilled/frozen, *reheated*	12 samples. 10-18% meat	76.9	1.49	9.3	5.7	5.3	108	454
274	**Stuffed cabbage leaves**	Recipe	78.6	1.12	7.0	4.4	8.0[d]	98	410
275	**Stuffed peppers**	Recipe	78.1	0.98	6.1	4.3	9.1[d]	97	408
276	**Sweet and sour pork**	Recipe	59.7	2.03	12.7	8.8	11.3[e]	179	958
277	-, made with lean pork	Recipe	61.6	2.11	13.2	7.0	11.3[e]	164	900
278	-, made with canned sweet and sour sauce	Recipe	69.8	2.19	13.7	6.6	5.4	135	564
279	**Tagliatelle with ham, mushroom and cheese**, chilled/frozen/longlife, *reheated*	11 samples, 9 brands. 10-14% meat	72.1	0.90	5.6	7.1	14.1[b]	139	584
280	**Toad in the hole**	Recipe	48.1	1.90	11.9	17.4	19.5	277	1158
281	-, made with skimmed milk and reduced fat sausages	Recipe	55.9	2.02	12.6	8.6	19.2	200	840
282	**Tripe and onions**, *stewed*	Recipe	78.2	1.33	8.3	2.7	9.5[b]	93	393
283	**Turkey and pasta bake**	Recipe	73.7	1.89	11.8	5.9	7.3[c]	128	536

[a] Includes 0.3g oligosaccharides per 100g food
[c] Includes 0.1g oligosaccharides per 100g food
[e] Includes 0.4g oligosaccharides per 100g food

[b] Includes 0.8g oligosaccharides per 100g food
[d] Includes 0.2g oligosaccharides per 100g food

Meat dishes *continued*

Composition of food per 100g

No. 19-	Food	Starch g	Total sugars g	Dietary fibre Southgate method g	Dietary fibre Englyst method g	Fatty acids Satd g	cis & trans Mono-unsatd g	cis & trans Poly-unsatd g	Total trans g	Cholesterol mg
269	**Sausage casserole**	2.7	2.1	1.4	0.9	3.5	4.5	2.0	0	40
270	**Shepherd's pie**	8.4	0.9	0.8	0.7	2.2	2.2	0.9	0.4	22
271	**Shish kebabs, with onions and peppers**	0.1	3.0	(1.4)	1.2	5.8	5.9	2.8	1.0	51
272	**Spaghetti bolognese**	11.0	1.4	1.2	0.9	2.1	2.2	0.5	0.2	17
273	-, chilled/frozen, *reheated*	2.3	3.0	N	0.9	2.3	2.5	0.5	0.2	N
274	**Stuffed cabbage leaves**	6.2	1.6	(1.2)	1.0	1.8	1.6	0.3	0.3	34
275	**Stuffed peppers**	6.4	2.5	(1.3)	1.0	1.8	1.8	0.2	0.2	15
276	**Sweet and sour pork**	3.4	7.5	0.6	0.6	2.0	3.3	2.8	Tr	50
277	-, made with lean pork	3.4	7.5	0.6	0.6	1.3	2.6	2.5	Tr	49
278	-, made with canned sweet and sour sauce	1.7	3.7	N	N	1.9	2.6	1.7	Tr	43
279	**Tagliatelle with ham, mushroom and cheese,** chilled/frozen/longlife, *reheated*	11.7	1.6	N	0.3	2.7	1.6	0.4	0.3	22
280	**Toad in the hole**	16.0	3.5	0.9	1.1	6.7	7.6	2.3	0.1	105
281	-, made with skimmed milk and reduced fat sausages	16.2	3.0	N	1.2	2.9	3.6	1.3	Tr	88
282	**Tripe and onions,** *stewed*	4.0	4.7	0.7	0.7	1.5	0.8	0.1	Tr	58
283	**Turkey and pasta bake**	5.1	2.1	(0.9)	0.6	2.7	1.8	0.9	Tr	35

Inorganic constituents per 100g food

No. 19-	Food	Na	K	Ca	Mg	P	Fe	Cu	Zn	Cl	Mn	Se	I
						mg						µg	
269	**Sausage casserole**	650	250	32	18	140	0.9	0.06	1.2	790	0.11	6	5
270	**Shepherd's pie**	340	230	16	13	69	0.7	0.06	1.0	340	0.07	1	4
271	**Shish kebabs, with onions and peppers**	200	330	27	24	140	1.6	0.09	2.6	300	0.18	(2)	6
272	**Spaghetti bolognese**	130	170	12	16	76	0.9	0.09	1.4	150	0.18	2	(4)
273	**-**, chilled/frozen, *reheated*	410	290	21	18	85	1.3	0.08	1.5	(630)	0.15	N	N
274	**Stuffed cabbage leaves**	210	210	25	1	81	1.0	0.05	1.2	210	0.14	(2)	5
275	**Stuffed peppers**	140	210	17	14	63	0.9	0.05	1.1	220	0.16	(2)	4
276	**Sweet and sour pork**	490	310	15	20	140	0.9	0.13	1.3	720	0.06	N	(6)
277	**-**, made with lean pork	490	310	15	21	150	0.9	0.13	1.4	720	0.06	N	(6)
278	**-**, made with canned sweet and sour sauce	240	280	1	18	140	0.8	0.06	1.4	270	0.11	8	N
279	**Tagliatelle with ham, mushroom and cheese**, chilled/frozen/longlife, *reheated*	340	110	69	14	110	0.5	0.11	0.6	660	0.17	N	N
280	**Toad in the hole**	670	200	130	17	200	1.2	0.08	1.1	1030	0.19	(6)	(19)
281	**-**, made with skimmed milk and reduced fat sausages	650	200	130	18	190	1.2	0.10	1.2	910	0.21	(6)	19
282	**Tripe and onions**, *stewed*	240	150	120	11	79	0.4	0.06	0.9	340	0.09	N	1
283	**Turkey and pasta bake**	240	210	70	18	140	0.6	0.14	1.1	370	0.10	7	(9)

Meat dishes *continued*

No. 19-	Food	Retinol µg	Carotene µg	Vitamin D µg	Vitamin E mg	Thiamin mg	Ribo-flavin mg	Niacin mg	Trypt 60 mg	Vitamin B6 mg	Vitamin B12 µg	Folate µg	Panto-thenate mg	Biotin µg	Vitamin C mg
269	Sausage casserole	Tr	19	0.5	N	0.32	0.10	2.7	2.1	0.21	Tr	5	0.60	2	Tr
270	Shepherd's pie	16	12	0.3	(0.34)	0.09	0.04	1.0	1.2	0.16	Tr	6	0.30	1	1
271	Shish kebabs, with onions and peppers	5	135	0.3	(0.41)	0.11	0.09	2.3	2.4	0.28	1	12	0.50	N	30
272	Spaghetti bolognese	Tr	240	0.1	(0.30)	0.04	0.04	1.7	1.4	0.13	Tr	6	0.17	1	2
273	-, chilled/frozen, *reheated*	Tr	N	N	N	N	N	N	N	N	N	N	N	N	Tr
274	Stuffed cabbage leaves	8	275	0.2	(0.15)	0.07	0.06	1.3	1.4	0.11	Tr	23	0.33	1	9
275	Stuffed peppers	Tr	170	N	N	0.05	0.04	1.4	1.1	0.21	Tr	12	0.19	N	26
276	Sweet and sour pork	7	450	0.4	(0.30)	0.46	0.13	3.4	2.6	0.30	1	6	0.72	2	14
277	-, made with lean pork	7	450	0.4	(0.30)	0.50	0.14	3.5	2.8	0.32	1	6	0.76	2	14
278	-, made with canned sweet and sour sauce	Tr	N	0.4	N	0.56	0.13	3.5	2.8	N	1	N	N	N	N
279	Tagliatelle with ham, mushroom and cheese, chilled/frozen/longlife, *reheated*	Tr	Tr	0.3	0.43	0.15	0.16	2.1	1.1	0.08	Tr	4	0.37	3	N
280	Toad in the hole	47	7	0.6	0.79	0.06	0.16	1.6	2.1	0.12	1	9	0.76	6	2
281	-, made with skimmed milk and reduced fat sausages	30	Tr	N	0.30	0.08	0.16	1.3	2.1	0.15	1	8	0.71	5	13
282	Tripe and onions, *stewed*	28	15	Tr	0.19	0.06	0.08	0.3	1.5	0.09	Tr	8	0.20	2	2
283	Turkey and pasta bake	35	390	0.1	0.15	0.07	0.13	2.7	2.4	0.18	Tr	8	0.50	(2)	7

Composition of food per 100g

No. Food 19-	Description and main data sources	Water g	Total Nitrogen g	Protein g	Fat g	Carbo-hydrate g	Energy value kcal	Energy value kJ
284 **Turkey, stir-fried with vegetables**	Recipe	79.7	1.81	11.3	2.8	3.7	84	355
285 **Venison in red wine and port**	Recipe	79.9	1.57	9.8	2.6	3.5[a]	76	319
286 **Wiener schnitzel**	Recipe	55.1	3.34	20.9	10.0	13.1	223	935

[a] Includes 0.2g oligosaccharides per 100g food

Meat dishes *continued*

No. Food	Starch	Total sugars	Dietary fibre Southgate method	Dietary fibre Englyst method	Fatty acids Satd	*cis & trans* Mono- unsatd	Poly- unsatd	Total trans	Cholesterol
19-	g	g	g	g	g	g	g	g	mg
284 **Turkey, stir-fried with vegetables**	1.5	2.2	N	1.1	0.5	0.9	1.1	Tr	30
285 **Venison in red wine and port**	1.7	1.6	0.4	0.4	1.5	0.6	0.2	Tr	26
286 **Wiener schnitzel**	12.7	0.4	0.9	0.4	1.6	3.6	3.9	Tr	81

No. Food						mg							µg	
19-	Na	K	Ca	Mg	P	Fe	Cu	Zn	Cl	Mn			Se	I
284 Turkey, stir-fried with vegetables	360	270	17	20	(120)	1.1	0.05	1.0	490	0.09			N	N
285 Venison in red wine and port	290	200	12	14	99	1.7	0.11	1.1	270	0.06			4	N
286 Wiener schnitzel	290	310	32	26	220	1.1	0.04	2.1	410	0.12			8	14

No. 19-	Food	Retinol µg	Carotene µg	Vitamin D µg	Vitamin E mg	Thiamin mg	Ribo-flavin mg	Niacin mg	Trypt 60 mg	Vitamin B6 mg	Vitamin B12 µg	Folate µg	Panto-thenate mg	Biotin µg	Vitamin C mg
284	Turkey, stir-fried with vegetables	Tr	825	0.1	N	0.05	0.10	3.1	2.2	0.29	1	16	N	N	20
285	Venison in red wine and port	N	N	N	N	N	0.09	N	N	N	N	N	N	N	0
286	Wiener schnitzel	20	Tr	1.2	(0.26)	0.11	0.19	5.1	4.5	0.43	1	14	0.73	3	0

Appendices

WEIGHT LOSSES ON COOKING MEAT PRODUCTS AND DISHES

Many meat products lose variable amounts of fat and juices when cooked, and these were discarded. Frozen, chilled and longlife products also lose water when reheated or cooked according to manufacturers' instructions. Each sample that was cooked or reheated was weighed both before and afterwards, and the losses are shown below. Where the sample was a composite of a number of individual products, no value can be given for the range of weight losses and a dash (–) is shown for the number of samples.

The weight losses found when recipe dishes were cooked is given in the Appendix on page 112.

No. 19-	Food	Number of samples	% loss Means and ranges
Bacon and ham			
2	**Bacon rashers, back**, *dry-fried*	9	33 (23 – 40)
3	-, *grilled*	15	32 (22 – 44)
4	-, *grilled crispy*	10	53 (28 – 68)
5	-, *microwaved*	15	39 (26 – 48)
6	-, dry cured, *grilled*	7	28 (20 – 36)
8	-, fat trimmed, *grilled*	15	33 (25 – 44)
–	-, rind on, *grilled*	10	32 (22 – 41)
9	-, reduced salt, *grilled*	6	32 (24 – 38)
10	-, smoked, *grilled*	10	32 (17 – 41)
11	-, sweetcure, *grilled*	9	29 (21 – 38)
12	-, 'tendersweet', *grilled*	10	31 (23 – 53)
14	**middle**, *fried*	9	34 (25 – 42)
15	-, grilled	7	38 (27 – 49)
17	**streaky**, *fried*	10	33 (24 – 44)
18	-, grilled	10	35 (21 – 51)
19	**Bacon loin steaks**, *grilled*	7	30 (23 – 37)
21	**Ham, gammon joint**, *boiled*	10	29 (15 – 40)
22	**Ham, gammon rashers**, *grilled*	5	34 (32 – 50)
Burgers and grillsteaks			
29	**Beefburgers**, chilled/frozen, *fried*	–	38
30	-, *grilled*	–	34
36	low fat, *fried*	–	22
37	-, *grilled*	–	20
43	**Economy burgers**, frozen, *grilled*	8	17 (11 – 23)

No. 19-	Food	Number of samples	% loss Means and ranges

Burgers and grillsteaks *continued*

No.	Food	Number of samples	% loss Means and ranges
45	**Grillsteaks, beef**, chilled/frozen, *fried*	–	32
46	-, *grilled*	–	25

Meat pies and pastries

No.	Food	Number of samples	% loss Means and ranges
51	**Beef pie**, chilled/frozen, *baked*	17	8 (6 – 10)
55	**Chicken pie**, individual, chilled/frozen, *baked*	11	5 (3 – 9)

Sausages

No.	Food	Number of samples	% loss Means and ranges
76	**Beef sausages**, chilled, *fried*	–	20
77	-, -, *grilled*	–	25
79	-, -, *fried*	–	17
80	**Pork sausages**, chilled, *grilled*	–	19
82	-, frozen, *fried*	–	20
83	-, *grilled*	–	24
85	reduced fat, chilled/frozen, *fried*	–	18
86	-, *grilled*	–	27
91	**Pork and beef economy sausages**, chilled, *fried*	–	19
92	-, *grilled*	–	28
94	**Premium sausages**, chilled, *fried*	–	21
95	-, *grilled*	–	24

Other meat products

No.	Food	Number of samples	% loss Means and ranges
114	**Black pudding**, *dry-fried*	–	12
116	**Chicken in crumbs**, stuffed with cheese and vegetables, chilled/frozen, *baked*	7	12 (7 – 20)
118	**Chicken breast in crumbs**, chilled, *fried*	–	5
119	-, *grilled*	–	13
120	**Chicken breast, marinated with garlic and herbs**, chilled, *baked*	6	15 (8 – 21)
122	**Chicken goujons**, chilled/frozen, *baked*	7	16 (7 – 28)
123	**Chicken kiev**, frozen, *baked*	–	13
127	**Chicken tandoori**, chilled, *baked*	–	18
	-, *microwaved*	–	13

No. 19-	Food	Number of samples	% loss Means and ranges

Other meat products *continued*

No. 19-	Food	Number of samples	% loss Means and ranges
131	**Faggots in gravy**, chilled/frozen,		
	baked	6	6 (4 – 10)
	-, *microwaved*	2	2 (2 – 3)
134	**Lamb roast**, frozen, *cooked*	–	32
147	**Pork roast**, frozen, *cooked*	–	37
155	**Turkey roast**, frozen, *cooked*	–	28

Meat dishes

No. 19-	Food	Number of samples	% loss Means and ranges
169	**Beef curry**, chilled/frozen, *baked*	–	5
	-, *microwaved*	–	16
172	**Beef in sauce with vegetables**, chilled/frozen, *reheated*	6	3 (1 – 4)
–	**Beef stew**, chilled/frozen/longlife		
	baked	3	14 (8 – 19)
	-, *microwaved*	3	5 (3 – 7)
188	**Chicken curry**, chilled/frozen		
	baked	2	1
	-, *microwaved*	–	24
193	**Chicken in sauce with vegetables**, chilled/frozen, *reheated*	8	7 (2 – 11)
201	**Chicken, stir-fried with rice and vegetables**, frozen, *reheated*	6	19 (15 – 23)
–	**Chicken tikka masala**, chilled/frozen		
	baked	3	3 (Tr – 4)
	-, *microwaved*	4	6 (2 – 9)
208	**Chilli con carne**, chilled/frozen, *reheated*	9	6 (0 – 18)
216	**Cottage/Shepherd's pie**, chilled/frozen		
	baked	9	13 (1 – 48)
	-, *microwaved*	2	21 (1 – 40)
231	**Lamb/Beef hot pot with potatoes**, chilled/frozen, *baked*	6	13 (6 – 21)
	-, *microwaved*	4	8 (1 – 14)
238	**Lasagne**, chilled/frozen	12	17 (7 – 47)
	baked	6	9 (7 – 10)
	-, *microwaved*	6	25 (8 – 47)
248	**Moussaka**, chilled/frozen/longlife		
	baked	4	9 (4 – 13)
	-, *microwaved*	4	7 (5 – 11)
249	**Pancakes, beef**, frozen, *shallow-fried*	–	6
250	**Pancakes, chicken**, frozen, *shallow-fried*	–	6
273	**Spaghetti bolognese**, chilled/frozen		
	baked	4	7 (6 – 8)
	-, *microwaved*	2	15 (11-19)

No. 19-	Food	Number of samples	% loss Means and ranges
	Meat dishes *continued*		
279	**Tagliatelle with ham, mushroom and cheese**, chilled/frozen/ longlife *baked*	4	3 (2 – 5)
	-, *microwaved*	7	9 (1 – 16)

RECILPES

Where possible, all recipes have been standardised to serve four people.

Where a recipe source indicated a portion but not the quantity of an ingredient the portion size was taken from Food Portion Sizes (MAFF, 1993) or weighed during recipe testing.

An egg was assumed to weigh 50g and a clove of garlic 4g. A level teaspoon refers to a standard 5ml spoon and was taken to hold 5g salt and 4g baking powder. The amounts of beer, lemon juice, milk, stock, vinegar, water and wine are given in millilitres but for beer, milk and wine, the millilitre measures were converted to gram weights for the purpose of recipe calculation. Stock was made up using 6g stock cube to 190ml water.

Quantities have not been included for recipes obtained in confidence from manufacturers. The ingredients have however, been listed in quantity order.

Unless specified, all the recipe items used were raw. Whole pasteurised milk, Cheddar cheese, plain white flour and distilled water were used. The bacon was derinded, the carrots, onions, potatoes and root ginger were peeled, the chilli peppers and peppers were deseeded, and except where otherwise specified, the turkey and chicken were skinless and boneless, and the beef, lamb and pork included both lean and fat.

Where canned fruit were used as ingredients, the nutrient profile was an average of the fruit canned in syrup and juice. Where canned tomatoes were used, the nutrient profile included the juice as well.

The type of fat used in the recipes has been specified. The vegetable oil was a retail blended vegetable oil. Margarine was an average of hard, soft and polyunsaturated types. The butter was salted. For fried dishes, the fat used during frying has been included at the end of the ingredients list with the quantity absorbed shown in brackets.

The baking powder used was a proprietary preparation whose composition is given in *Miscellaneous Foods* (Chan *et al*, 1994). Use of another brand could result in a different composition in the cooked dish with respect to sodium, calcium and phosphorus.

The nutrients in the small amounts of herbs and spices used in the recipes were ignored for the purpose of recipe calculations.

Recipes of ingredients used within the main recipes

Flaky pastry

200g flour	½tsp salt
75g margarine	85ml water
75g lard	10ml lemon juice

Divide fat into four portions. Sift flour and salt, rub in one portion of fat. Mix with water and lemon juice, then knead until smooth and leave for 15 minutes. Roll out, dot two-thirds with another fat portion and fold into 3. Roll out and repeat process with remaining 2 fat portions.

Shortcrust pastry

200g flour	½tsp salt
50g margarine	30ml water
50g lard	

Rub the fat into the flour, add salt, mix to a stiff dough with the water.

White sauce

25g margarine	25g flour
350ml whole or semi-skimmed milk	½tsp salt

Melt margarine, add flour and cook for a few minutes, stirring constantly. Add milk and salt and cook gently until mixture thickens.

Weight loss: 18%

Main recipes

31 Beefburgers, homemade, fried

500g minced beef	½tsp salt
60g onions, finely chopped	¼tsp pepper
1 egg	1tbsp vegetable oil

Combine the minced beef, onions, egg and seasoning. Divide into four and shape into patties. Fry in oil for 2 minutes each side on a high heat, then 6 minutes on low heat.

Weight loss: 29%

32 Beefburgers, homemade, grilled

As fried beefburgers (No 31) except omit oil and grill for 6 minutes each side.

Weight loss: 31%

33 Beefburgers, homemade, fried, with bun

210g fried beefburgers, homemade	50g hamburger bun

Proportions only

34 **Beefburgers, homemade, grilled, with bun**

210g grilled beefburgers, homemade   50g hamburger bun

Proportions only

52 **Beef steak pudding**

Suet crust pastry:	*Filling:*
200g flour	500g braising steak, diced
100g suet	50g flour
1½tsp baking powder	1tsp salt
½tsp salt	¼tsp pepper
130ml water	130g onions, chopped
	25ml water

Make the suet crust pastry, and line a pudding basin leaving enough for a lid, add in the meat rolled in seasoned flour, and onions. Add water and cover with the remaining pastry. Steam for 3 hours.

Weight gain: 2%

54 **Chicken and mushroom pie, single crust, homemade**

15g butter	80g mushrooms, sliced
15g flour	½tsp salt
255ml milk	¼tsp pepper
250g chicken light meat, roasted, diced	150g shortcrust pastry
40g boiled potatoes, diced	

Glaze:
5ml milk

Cook the butter and flour for 1 minute, stir in the milk. Bring to the boil and cook for a few minutes. Add the chicken, potatoes, mushrooms, seasoning and mix. Transfer to a pie dish and cover with pastry. Brush with milk and bake for 25-30 minutes at 200°C/mark 6.

Weight loss: 13%

57 **Cornish pastie**

2tsp vegetable oil	¼tsp pepper
100g minced beef	300g shortcrust pastry
100g swede, diced	10ml water
100g potatoes, diced	
50g carrots, diced	
30g onions, chopped	*Glaze:*
¼tsp salt	1tbsp beaten egg

Brown the mince in oil. Add the vegetables, seasoning and cook for 10 minutes stirring occasionally. Divide the pastry into four and roll each piece to a 20cm circle, add a quarter of the cooked meat mixture to each. Fold the edges up to the centre and pinch to seal with a little water. Brush with beaten egg, and bake for 40 minutes at 180°C/mark 4.

Weight loss: 14%

58 Game pie

shortcrust pastry
venison
middle bacon rashers
rabbit
chicken liver

pork spare-rib
pheasant
redcurrant jelly
onions
red wine

Proportions of main ingredients obtained from manufacturers.

60 Lamb samosas, homemade, baked

Filling:
225g minced lamb
100g onions, finely chopped
2 cloves garlic, crushed
1tbsp vegetable oil
½tsp chilli powder
½tsp garam masala
2tbsp fresh coriander leaves,
   chopped

100g boiled potatoes, diced
60g peas, boiled
½tsp salt
¼tsp pepper

Pastry:
125g flour
60ml water
2tsp vegetable oil

To bake:
4tbsp vegetable oil

Make the pastry. Brown the meat, onions and garlic in oil, add the spices and then the other ingredients and cook gently until soft. Divide the pastry into 8 pieces and roll out into circles 13cm in diameter. Cut each in half, add filling and fold to form a cone. Seal edges with a little water and brush each with oil and bake for 15-20 minutes at 180°C/mark 4, turning occasionally.

Weight loss: 25%

61 Lamb samosas, deep-fried, homemade

As baked lamb samosas (No 60), except deep-fried in oil.

(190g vegetable oil)

Weight loss: 14%

67 Sausage rolls, flaky pastry, homemade

225g pork sausage meat
10g flour

175g flaky pastry

Glaze:
10ml milk

Roll the pastry on a floured surface and cut into 10cm squares. Divide the sausage meat equally and place in the centre of each square. Fold over, seal, glaze with a little milk and bake for 20-30 minutes at 220°C/mark 7.

Weight loss: 14%

68 Sausage rolls, short pastry, homemade

225g pork sausage meat
10g flour

175g shortcrust pastry

Glaze:
10ml milk

Roll the pastry on a floured surface and cut into 10cm squares. Divide the sausage meat equally and place in the centre of each square. Fold over, seal, glaze with a little milk and bake for 20-30 minutes at 220°C/mark 7.

Weight loss: 8%

70 Steak and kidney pie, single crust, homemade

400g stewing beef, diced
200g lamb's kidneys, diced
2tsp salt

15g flour
100ml water
350g flaky pastry

Place the meat and kidneys rolled in seasoned flour in a pie dish with the water. Cover with pastry. Bake for 20 minutes at 200°C/mark 6, then lower the heat to 150°C/mark 2 and cover with greaseproof paper. Cook for 2-2½ hours more.

Weight loss: 21%

71 Steak and kidney pie, double crust, homemade

400g stewing beef, diced
200g lamb's kidneys, diced
100ml water

2tsp salt
15g flour
700g flaky pastry

Place the meat and kidneys rolled in seasoned flour in a pie dish lined with pastry, cover with additional pastry. Bake for 20 minutes at 200°C/mark 6, then lower the heat to 150°/mark 2 and cover with greaseproof paper. Cook for 2-2½ hours more.

Weight loss: 21%

73 Steak and kidney pudding, homemade

Suet crust pastry:
200g flour
100g suet
1½tsp baking powder
½tsp salt
130ml water

Filling:
400g braising steak, diced
100g lamb's kidneys, diced
50g flour
1tsp salt
¼tsp pepper
130g onions, chopped
25ml water

Make the suet crust pastry, and line a pudding basin. Put the meat and kidneys rolled in seasoned flour into the basin with the onions. Add the water and cover with the remaining pastry. Steam for 3 hours.

Weight gain: 2%

74 Turkey pie, single crust, homemade

15g butter
15g flour
255ml milk
250g cooked light and dark turkey
  meat, diced
50g peas, boiled
50g carrots, boiled, diced

50g potatoes, boiled, diced
½tsp salt
¼tsp pepper
150g shortcrust pastry

Glaze
5ml milk

Cook the butter and flour for 1 minute, stir in the milk. Bring to the boil, stirring until the sauce thickens. Add the turkey, vegetables and seasoning. Transfer to a pie dish and cover with pastry. Brush with milk and bake for 25-30 minutes at 200°C/mark 6.

Weight loss: 13%

101 Frankfurter with bun

23g frankfurter

50g hotdog bun

Proportions only

102 Frankfurter with bun, ketchup, fried onions and mustard

23g frankfurter
40g fried onions, chopped
5g mustard

50g hotdog bun
12g tomato ketchup

Proportions only

137 Meat loaf, homemade

1tsp vegetable oil for greasing
250g minced beef
250g minced pork
1tbsp Worcestershire sauce
1tbsp dried mixed herbs

1 egg
100g fresh breadcrumbs
100g onions, finely chopped
½tsp salt
¼tsp pepper

Grease a 500g loaf tin with oil. Mix the other ingredients. Press down well into the tin and bake for 1 hour at 180°C/mark 4.

Weight loss: 10%

160 Beef and spinach curry

1tbsp vegetable oil
225g onions, chopped
2 cloves garlic, crushed
500g stewing beef, diced
8g root ginger, grated
1tbsp coriander seeds, crushed
1tbsp ground cumin
1tsp ground turmeric

8 green cardamom pods,
  seeds removed, crushed
2 green chillies, chopped
397g canned tomatoes
200ml water
175g spinach, frozen
1tsp salt
¼tsp pepper

Brown the onions, garlic and meat in oil, add the ginger, spices and chillies. Stir in the tomatoes, water, spinach and seasoning. Bring to the boil, cover and simmer for 1 hour, stirring occasionally. Remove the lid and simmer for a further 45 minutes.

Weight loss: 28%

161 Beef bourguignonne

1tbsp vegetable oil	5g tomato purée
100g button onions	1tsp dried mixed herbs
1 clove garlic, crushed	250ml red wine
500g stewing beef, diced	250ml stock
50g streaky bacon rashers,	½tsp salt
chopped	¼tsp pepper
15g flour	150g button mushrooms

Brown the onions, garlic, meat and bacon in oil. Stir in flour, tomato purée, mixed herbs, wine, stock and seasoning. Bring to the boil, cover and simmer for 1 hour, stirring occasionally. Add mushrooms and cook for a further 30 minutes.

Weight loss: 33%

162 Beef bourguignonne, made with lean beef

As for beef bourguignonne (No 161), except made with lean stewing steak and back bacon rashers.

164 Beef casserole, made with canned cook-in sauce

500g braising steak, diced	390g cook-in sauce, canned

Cook the meat with sauce in a covered casserole dish for 1½ hours at 180˚C/ mark 4.

Weight loss: 20%

166 Beef curry

1 clove garlic, crushed	½tsp ground turmeric
60g onions, chopped	8g root ginger, grated
500g braising steak, diced	300ml water
6tbsp vegetable oil	½tsp salt
1tbsp ground coriander	5ml lemon juice
1tsp chilli powder	1tsp garam masala
½tsp ground cumin	

Brown the garlic, onions and meat in oil. Add spices and ginger. Stir in water, salt and lemon juice, cover and bring to the boil. Cook for 1½ hours stirring occasionally. Add garam masala.

Weight loss: 34%

167 Beef curry, reduced fat

As for beef curry (No 166), except made with 1tbsp vegetable oil and lean braising steak.

171 Beef enchiladas

200g tortillas	50g cheese, grated
1200g chilli con carne, homemade	

Fill tortillas with chilli con carne (No 206) and cheese.

173 Beef kheema

75g onions, finely chopped
2 garlic cloves, crushed
500g minced beef
2tbsp vegetable oil
8g root ginger, grated
2 green chillies, finely chopped
1tsp coriander seeds, crushed
1tsp ground cumin
1tsp cayenne pepper

200ml water
200g peas, frozen
2tbsp fresh coriander
  leaves, chopped
1tsp salt
¼tsp pepper
2tsp garam masala
220g canned tomatoes

Brown the onions, garlic and mince in oil. Add ginger and spices. Stir in 150ml of the water, cover and simmer for 30 minutes. Add the peas, coriander leaves, seasoning, garam masala, tomatoes and the remaining water and bring back to the boil. Cover and cook a further 10 minutes.

Weight loss: 21%

174 Beef olives

500g beef topside,
  cut into 8 thin slices
½tsp salt
¼tsp pepper
200g back bacon rashers,
  chopped

1tsp dried mixed herbs
50g onions, finely chopped
1tbsp vegetable oil
15g flour
400ml stock

Season the beef, put one rasher of bacon on each slice and top with onions and herbs. Roll and secure with string. Brown in oil, add flour and stock, cover and cook in the oven at 170°C/mark 3 for 1 hour.

Weight loss: 16%

175 Beef stew

500g stewing beef, diced
150g onions, chopped
1tbsp vegetable oil
30g flour

500ml stock
150g carrots, chopped
½tsp salt
¼tsp pepper

Brown the meat and onions in oil, add flour and cook for 1 minute. Blend in the stock, add carrots and seasoning, transfer to a dish, cover, and cook in the oven for 2 hours at 180°C/mark 4.

Weight loss: 27%

176 Beef stew, made with lean beef

As for beef stew (No 175), except made with lean stewing beef.

177 Beef stew and dumplings

Dumplings:
200g self-raising flour
100g suet

¼tsp salt
120ml water

As for beef stew (No 175). Combine dumpling ingredients. Divide this mixture into walnut-sized pieces and roll them into balls. Add to the stew for the last 30 minutes of the cooking time.

Weight loss: 24% (whole dish)

180 Beef, stir-fried with green peppers

500g rump steak, thinly sliced
2tbsp vegetable oil
400g green peppers, sliced
60g spring onions, sliced
20g root ginger, grated

Marinade:
4tsp sugar
1 red chilli, finely chopped
2tbsp soy sauce
2tbsp sherry
20g cornflour
¼tsp salt
¼tsp pepper

Marinade the steak for 30 minutes. Stir-fry the peppers, spring onions and ginger in oil for a few minutes, then add meat and stir-fry for 6 minutes.

Weight loss: 16%

181 Beef Stroganoff

15g butter
150g onions, sliced
500g fillet steak, strips
150g mushrooms, sliced

½tsp salt
¼tsp pepper
150g soured cream

Brown the onions and steak in butter, add mushrooms and cook for 10-12 minutes. Add seasoning and soured cream, and heat through without boiling.

Weight loss: 24%

182 Beef Wellington

400g beef fillet, tied with
  string
½tsp salt
¼tsp pepper
1tbsp vegetable oil

100g smooth liver pâté
10g flour
300g puff pastry

Glaze:
beaten egg

Season the fillet and brown in oil. Roast for 20 minutes at 220°C/mark 7, allow to cool and remove string. Roll out the pastry on a floured surface to a large rectangle 0.5cm thick, spread the pâté mixture down the centre and top with the meat. Brush the edges of the pastry with egg. Fold edges over lengthways and turn so that the join is underneath. Decorate with remaining pastry trimmings and bake for 45 minutes at 220°C/mark 7, covering with foil after 25 minutes.

Weight loss: 17%

183 Bolognese sauce

1 clove garlic, crushed
60g onions, chopped
500g minced beef
2tsp vegetable oil
40g carrots, chopped
30g celery, chopped
10g tomato purée

397g canned tomatoes
125ml stock
125ml red wine
½tsp salt
¼tsp pepper
¼tsp dried mixed herbs

Brown the garlic, onions and mince in oil, add carrots and celery. Stir in the other ingredients and simmer for 40 minutes with the lid on.

Weight loss: 32% (whole dish)

185 **Carbonnade of beef**

300g onions, sliced
100g streaky bacon rashers, chopped
500g stewing beef, diced
1tbsp vegetable oil
300ml beer
150ml stock

25g flour
½tsp salt
¼tsp pepper
5g made mustard
1tsp brown sugar
¼tsp dried mixed herbs

Brown the onions, bacon and meat in oil. Blend in the beer, flour, and stock, and boil until thickened. Add remaining ingredients and cook in a covered casserole for 2 hours at 150°C/mark 2.

Weight loss: 25%

186/7 **Chicken chasseur**

800g chicken breast (weighed with bone)
150g shallots
1tbsp vegetable oil
1tbsp flour
300ml dry white wine
300ml stock
2 bay leaves

1tsp dried mixed herbs
½tsp salt
¼tsp pepper
1 clove garlic, crushed
15g tomato purée
1tsp brown sugar
100g button mushrooms

Brown the chicken and shallots in oil. Remove and transfer to a casserole dish. Add the flour to the pan and gradually blend in the wine and stock and bring to the boil. Add remaining ingredients and stir. Pour over the chicken, cover and cook for 1 hour at 180°C/mark 4.

Weight loss: 17% (with bone), 18% (without bone)

190 **Chicken curry, made with canned curry sauce**

1tbsp vegetable oil
500g chicken breast

385g curry sauce, canned

Brown the chicken in oil. Add sauce, cover and simmer for 45 minutes.

Weight loss: 30%

191 **Chicken fricassée**

225g back bacon rashers, chopped
450g shallots, sliced
600g chicken breast, chopped
15g butter
2tbsp vegetable oil
30g flour
300ml white wine

300ml stock
2 bay leaves
½tsp salt
¼tsp pepper
150g button mushrooms
15g smooth mustard
150g single cream

Brown the bacon, shallots and chicken in butter and oil, add flour, white wine, stock, bay leaves and seasoning, cover and simmer for 25 minutes. Add the mushrooms and cook for a further 20 minutes. Stir in the mustard and cream.

Weight loss: 16%

192 **Chicken fricassée, reduced fat**

As for chicken fricassée (No 191), except omit butter and replace cream with plain yogurt.

Weight loss: 16%

195 Chicken in white sauce, made with whole milk

275g cooked light and dark chicken meat, diced

225g white sauce, made with whole milk

Mix the ingredients together.

196 Chicken in white sauce, made with semi-skimmed milk

275g cooked light and dark chicken meat, diced

225g white sauce, made with semi-skimmed milk

Mix the ingredients together.

197 Chicken korma

150g onions, finely chopped
2 cloves garlic, crushed
500g chicken breast, diced
2tbsp vegetable oil
15g flour
1tbsp coriander seeds, crushed
1tbsp desiccated coconut
1tsp ground turmeric

½tsp ground cumin
½tsp chilli powder
2tsp ground almonds
15ml lemon juice
½tsp salt
¼tsp pepper
300ml milk
1tbsp plain yogurt

Brown the onions, garlic and chicken in oil. Add the other ingredients except the milk and yogurt, and cook for a few minutes. Stir in the milk and simmer for 40 minutes. Add in the yogurt.

Weight loss: 16%

198 Chicken risotto

140g onions, finely chopped
1 clove garlic, crushed
250g chicken breast, diced
20g butter
300g glutinous rice, raw
600ml stock

½tsp salt
¼tsp pepper
1tbsp fresh parsley, finely chopped
10g Parmesan cheese, grated

Brown the onions, garlic and chicken in butter. Add rice and cook for a few minutes, stirring constantly. Add in the stock and simmer for 20 minutes stirring occasionally. Add seasoning, parsley and Parmesan.

Weight loss: 30%

199 Chicken, stir-fried with mushrooms and cashew nuts

5g root ginger, grated
1 clove garlic, crushed
500g chicken breast, strips
1tbsp vegetable oil
150g red peppers, sliced
50g button mushrooms, halved

50g cashew nuts, roasted
2tbsp dark soy sauce
2tbsp sherry
10g cornflour
¼tsp black pepper

Stir-fry the ginger, garlic and chicken in oil until brown. Add peppers, mushrooms and cashew nuts and stir-fry for a few minutes. Add soy sauce, sherry, cornflour and black pepper, stirring continuously, until meat and vegetables are coated.

Weight loss: 19%

200 Chicken, stir-fried with peppers in black bean sauce

400g chicken breast, strips
1tbsp vegetable oil
1 red chilli, thinly sliced
60g onions, chopped
50g green peppers, chopped
60g black bean sauce

Stir-fry the chicken in oil until brown. Add the chilli, onions, peppers and black bean sauce and stir-fry for 5-7 minutes.

Weight loss: 15%

202 Chicken vindaloo

8g root ginger, grated
3 cloves garlic, crushed
2 red chillies, finely chopped
100g onions, finely chopped
2tsp ground cumin seeds
1tsp black peppercorns, crushed
1tsp ground cardamom
1tsp ground cinnamon
1½tsp black mustard seeds, crushed
1tsp fenugreek seeds, crushed
1tbsp ground coriander
½tsp ground turmeric
500g chicken breast, diced
6tbsp vegetable oil
75ml vinegar
½tsp salt
1tsp sugar
200ml water

Brown the ginger, garlic, chilli, onions, spices and chicken in oil, add the remaining ingredients, cover, and simmer gently for 1½ hours stirring occasionally.

Weight loss: 36%

203 Chicken vindaloo, reduced fat

As for chicken vindaloo (No 202), except made with 1tbsp vegetable oil.

206 Chilli con carne

500g minced beef
150g onions, sliced
100g green peppers, chopped
1tbsp vegetable oil
1tsp salt
¼tsp pepper
2tsp chilli powder
15ml vinegar
1tsp sugar
30g tomato purée
397g canned tomatoes
150ml stock
115g red kidney beans, canned, drained

Brown the mince, onions and peppers in oil. Blend the other ingredients and stir into the meat. Cover and simmer gently for 40 minutes. Add the kidney beans and continue cooking for a further 10 minutes.

Weight loss: 15%

210/1 Coq au vin

100g back bacon rashers, chopped
1000g chicken leg quarters (weighed with bone)
50g flour
½tsp salt
¼tsp pepper
50g butter
100g shallots
1tsp dried mixed herbs
600ml red wine
100g button mushrooms

Brown the bacon and chicken coated in seasoned flour, in butter. Add the shallots, mixed herbs and red wine, cover and simmer for 35-45 minutes. Add the mushrooms and cook for another 20 minutes.

Weight loss: 16% (with bone), 19% (without bone)

212 **Corned beef hash**

200g onions, sliced
30g butter
450g corned beef, canned
800g boiled potatoes

30ml milk
½tsp salt
¼tsp pepper

Brown the onions in butter, add corned beef. Mash the potatoes with the milk. Mix together with seasoning and cook for 15 minutes.

Weight loss: 15%

213 **Coronation chicken**

300g mayonnaise
1tbsp curry paste
2tbsp apricot jam

500g cooked light and dark
  chicken meat, diced

Mix the ingredients together.

214 **Coronation chicken, reduced fat**

300g mayonnaise, reduced calorie
1tbsp curry paste
2tbsp apricot jam, reduced sugar

500g cooked light and dark
  chicken meat, diced

Mix the ingredients together.

215 **Cottage pie**

500g stewed minced beef
  (No 242)
100g carrots, sliced
½tsp salt

¼tsp pepper
500g boiled potatoes
50ml milk
20g margarine

Place the mince, carrots and seasoning in a pie dish. Mash the potatoes with the milk and margarine. Pile on top of the meat and bake for 1 hour at 190°C/mark 5.

Weight loss: 11%

217 **Devilled kidneys**

1tbsp vegetable oil
400g lamb's kidneys, skinned,
  cored, halved
100g mushrooms, sliced
1tbsp Worcestershire sauce

15g tomato purée
1tsp mustard
½tsp salt
¼tsp pepper
100ml single cream

Brown the kidneys and mushrooms in oil. Add Worcestershire sauce, tomato purée, mustard and seasoning and mix well. Cover and simmer gently for 10-15 minutes. Add cream and heat through without boiling.

Weight loss: 17%

218/9 Duck à l'orange

600g roast duck, meat, fat and skin
or 500g roast duck, meat only

Orange sauce:

25g butter	½tsp salt
35g cornflour	¼tsp pepper
450ml stock	30ml sherry
zest of 1 orange	100ml orange juice

Melt the butter, add cornflour, stock and orange zest and boil, stirring continuously until thickened. Add seasoning, sherry and orange juice, heat through. Serve over duck.

Weight loss: 5% (sauce only)

220 Duck with pineapple

600g roast duck, meat, fat and skin, strips	15g sugar
	20g spring onions, chopped
2 cloves garlic, crushed	140g canned pineapple, drained
10g root ginger, grated	
100g onions, chopped	150ml stock
4tsp soy sauce	¼tsp pepper
20ml sherry	10g cornflour

Stir-fry the duck, ginger, garlic and onions for a few minutes. Add remaining ingredients and cook for 5 minutes.

Weight loss: 18%

221 Goulash

300g onions, chopped	397g canned tomatoes
500g stewing beef, diced	150g green peppers, chopped
2tbsp vegetable oil	
2 cloves garlic, crushed	1tsp salt
2tsp paprika	1 litre stock
2g caraway seeds, crushed	500g potatoes, diced

Brown the onions and meat in oil. Add remaining ingredients except for the potatoes. Cover and simmer for 1 hour. Add potatoes and simmer for a further 25 minutes.

Weight loss: 32%

222 Irish stew

500g lamb neck fillet, diced	1tsp dried mixed herbs
150g onions, sliced	15g flour
200g carrots, sliced	½tsp salt
500g potatoes, sliced	¼tsp pepper
1tbsp fresh parsley, chopped	300ml stock

Make layers of meat, vegetables, herbs, flour, and seasoning in a casserole dish, ending with a top layer of potatoes. Pour in stock and cover. Bake for 1 hour at 170°C/mark 3, remove lid and cook for a further 30 minutes.

Weight loss: 13%

223 Irish stew, made with lean lamb

As for Irish stew (No 222), except made with lean lamb neck fillet.

225 Lamb biryani

400g stewing lamb, diced
8g root ginger, grated
2 cloves garlic, crushed
400g onions, sliced
1tsp cardamom seeds, crushed
11tbsp vegetable oil
400g white long grain rice
2tbsp garam masala
100g brown lentils, boiled
650ml water

Marinade:
1 green chilli, chopped
150g canned tomatoes
2tsp fresh parsley, chopped
1tsp turmeric
130g yogurt
½tsp salt

Marinade the meat for 1 hour. Brown the meat, ginger, garlic, onions and cardamom in oil, add rice, garam masala, lentils and water and simmer for 10 minutes before transferring the mixture to an ovenproof dish. Cover and bake for 30 minutes at 170°C/mark 3.

Weight loss: 17%

226 Lamb biryani, reduced fat

As for lamb biryani (No 225), except made with 1tbsp vegetable oil and low fat yogurt.

227 Lamb curry, made with canned curry sauce

500g stewing lamb, diced
1tbsp vegetable oil

385g curry sauce, canned

Brown the lamb in oil. Add sauce, cover and simmer for 45 minutes.

Weight loss: 30%

228 Lamb kheema

6tbsp vegetable oil
75g onions, finely chopped
2 garlic cloves, crushed
500g minced lamb
8g root ginger, grated
2 green chillies, deseeded,
  finely chopped
1tsp coriander seeds, crushed
1tsp ground cumin

1tsp cayenne pepper
200ml water
200g peas, frozen
2tbsp fresh coriander
  leaves, chopped
1tsp salt
2tsp garam masala
220g canned tomatoes

Brown the onions, garlic and mince in oil. Add the ginger and spices. Stir in 150ml of the water, cover and simmer for 30 minutes. Add the remaining ingredients and bring back to the boil. Cover and cook for a further 10 minutes.

Weight loss: 21%

229 Lamb kheema, reduced fat

As for lamb kheema (No 228), except made with 2tbsp vegetable oil.

230 Lamb koftas

500g minced lamb
1 green chilli, chopped
¼tsp salt
1tsp ground cumin seeds
50g fresh breadcrumbs
50g onion, chopped

¼tsp cayenne pepper
2tbsp fresh coriander
    leaves, finely chopped
1 egg
1tsp garam masala

Combine all the ingredients. Shape into 30 meatballs. Grill under a high heat for 15 minutes.

Weight loss: 32%

232 Lamb rogan josh

1tbsp vegetable oil
500g stewing lamb, diced
250g onions, finely chopped
2 cloves garlic, crushed
8g root ginger, grated
¼tsp ground cinnamon
1tsp ground coriander

2tsp ground cumin
4tsp paprika
2 cardamom pods, crushed
½tsp salt
¼tsp pepper
397g canned tomatoes
100ml water

Brown the meat, onions, garlic and ginger in oil, add spices and seasoning. Stir in tomatoes and water. Cover and simmer for 1 hour. Remove lid and simmer for a further 30 minutes.

Weight loss: 34%

233 Lamb, stir-fried with spring onions

500g lamb neck fillet, sliced
1tbsp vegetable oil
2 cloves garlic, chopped
225g spring onions, strips
2tsp sesame oil
¼tsp salt
½tsp sugar

Marinade:
4tsp soy sauce
4tsp dry sherry

Marinade the lamb for 20 minutes. Stir-fry the garlic, lamb and spring onions in oil until brown. Add sesame oil, salt and sugar, cook for a few minutes.

Weight loss: 25%

234 Lamb vindaloo

8g root ginger, grated
3 cloves garlic, crushed
2 red chillies, finely chopped
100g onions, finely chopped
2tsp cumin seeds, crushed
1tsp black peppercorns, crushed
1tsp ground cardamom
1tsp ground cinnamon
1½tsp black mustard seeds, crushed
1tsp fenugreek seeds, crushed

1tbsp ground coriander
½tsp ground turmeric
500g stewing lamb, diced
1tbsp vegetable oil
75ml vinegar
½tsp salt
1tsp sugar
200ml water

Brown the ginger, garlic, chilli, onions, spices and lamb in oil, add vinegar, salt and sugar. Add the water, cover, and simmer gently for 1½ hours stirring occasionally.

Weight loss: 36%

235 Lamb's heart casserole

30g butter
50g onions, chopped
50g mushrooms, sliced
50g streaky bacon rashers, chopped
1tsp dried sage
100g fresh breadcrumbs
10ml lemon juice
½tsp salt

¼tsp pepper
½ egg
680g lamb's hearts,
  washed and trimmed
50g flour
1tbsp vegetable oil
300ml stock
50ml sherry

Brown the onions, mushrooms and bacon in 15g butter. Stir in sage, breadcrumbs, lemon juice, seasoning and egg. Fill the hearts with the prepared stuffing and sew up with cotton. Coat in flour and brown in oil and remaining butter. Add stock and sherry. Cover and bake for 2 hours at 150°C/mark 2.

Weight loss: 12%

236 Lancashire hotpot

500g stewing lamb, diced
½tsp salt
¼tsp pepper
100g carrots, sliced
100g turnip, chopped

100g onions, sliced
500g potatoes, sliced
300ml stock
2tsp vegetable oil

Season the meat and mix with carrots, turnip and onions. Layer this with the potatoes in a casserole, beginning and ending with potatoes. Add stock and brush the top with oil. Cover and bake for 2 hours at 150°C/mark 2. Remove lid to brown the potatoes for the last 30 minutes.

Weight loss: 11%

237 Lasagne

Meat sauce:
1tbsp vegetable oil
50g streaky bacon rashers,
  chopped
50g onions, chopped
50g carrots, chopped
30g celery, chopped
300g minced beef
220g canned tomatoes
375ml stock
1 clove garlic, crushed
½tsp salt
¼tsp pepper
½tsp marjoram
1 bay leaf
50g mushrooms, sliced

Cheese sauce:
30g margarine
30g flour
400ml milk
75g cheese, grated

200g lasagne, raw

To top:
25g cheese, grated

Brown the bacon, onions, carrots, celery and mince in oil. Stir in the remaining ingredients for the meat sauce and simmer for 15 minutes. For the cheese sauce, melt the margarine, add flour and cook for a few minutes, stir in milk and cheese and cook gently until mixture thickens. In a dish, add alternative layers of lasagne, meat and cheese sauce ending with a layer of lasagne and cheese sauce. Sprinkle with cheese and bake for 1 hour at 190°C/mark 5.

Weight loss: 26%

239 Lemon chicken

500g chicken breast, strips
1 egg white
½tsp salt
1tbsp cornflour

Lemon sauce:
1tbsp cornflour
200ml water
1½tbsp sugar
60ml lemon juice

(40g vegetable oil)

Coat the chicken in egg white and salted cornflour then deep-fry for 3 minutes and drain. Prepare the sauce by blending the cornflour with water, add remaining ingredients and boil until clear and thickened. Add chicken and heat through.

Weight loss: 17%

240 Liver and bacon, fried

400g fried lamb's liver
½tsp salt
¼tsp pepper

100g fried back bacon
rashers

Proportions only.

241 Liver and onions, stewed

1tbsp vegetable oil
300g onions, chopped
300g lamb's liver, strips

½tsp salt
¼tsp pepper

Brown the onions and liver in oil, add seasoning and cook for 10 minutes, stirring continuously.

Weight loss: 29%

242 Minced beef, stewed

1tbsp vegetable oil
100g onions, chopped
500g minced beef
20g flour

500ml stock
¼tsp salt
¼tsp pepper

Brown the onions and mince in oil. Add flour and cook for a few minutes, stir in stock and seasoning and simmer gently for 30 minutes.

Weight loss: 17%

243 Minced beef, extra lean, stewed

As for stewed minced beef (No 242), except made with extra lean minced beef.

245 Minced beef with vegetables, stewed

200g onions, chopped
500g minced beef
200g carrots, chopped
550ml stock

45g flour
½tsp salt
75ml water

Brown the onions and mince. Add carrots and stock, cover and simmer for 45 minutes. Blend flour, salt and water together and stir into the minced beef. Bring to the boil and simmer for a further 10 minutes.

Weight loss: 26%

246 Minced lamb, stewed

100g onions, chopped
500g minced lamb
1 tbsp oil
500ml stock

20g flour
¼tsp salt
¼tsp pepper

Brown the onions and mince in oil. Add stock, flour and seasoning, cover and simmer for 45 minutes.

Weight loss: 17%

247 Moussaka

1 tsp vegetable oil
500g aubergines, sliced
150g onions, chopped
500g minced lamb
397g canned tomatoes
1 tsp mixed herbs
½tsp salt
¼tsp pepper

Cheese sauce:
30g margarine
30g flour
300ml milk
50g cheese, grated
½tsp salt
¼tsp pepper
1 egg

Fry the aubergines in oil for 2 minutes each side and remove from pan. Brown the onions and mince. Add the tomatoes, mixed herbs and seasoning. Bring to the boil and simmer uncovered for 15 minutes. For the cheese sauce, melt the margarine, add flour and cook for a few minutes, blend in milk and bring to the boil, stirring constantly. Add cheese and seasoning and allow to cool before beating in the egg. Arrange layers of aubergines and meat in an ovenproof dish. Pour on the cheese sauce and cook for 30 minutes at 190°C/mark 5.

Weight loss: 20%

251 Pasta, with ham and mushroom sauce

900g white pasta sauce
  with ham and mushrooms

900g boiled pasta

Mix ingredients together.

252 Pasta with meat and tomato sauce

340g minced beef
900g boiled pasta

475g pasta sauce,
  tomato-based, canned

Brown the mince in a pan. Add pasta sauce and simmer for 20 minutes. Stir in pasta.

Weight loss 17%: (meat sauce)

253 Pork and apple casserole

1 tbsp vegetable oil
500g pork steaks
100g onions, sliced
40g flour
300ml stock
200ml apple juice

½tsp salt
¼tsp pepper
1 tsp sage
150g eating apples, cored
  and sliced
100ml cider

Brown the pork and onions in oil. Add flour and cook for a few minutes. Blend in the stock, apple juice, seasoning and sage, bring to the boil stirring occasionally. Cover and bake for 1 hour at 180°C/mark 4. Add apple slices and cider and cook for a further 30 minutes.

Weight loss: 23%

254 Pork and beef meatballs in tomato sauce

250g minced pork
250g minced beef
25g fresh breadcrumbs
25g Parmesan cheese, grated
1tbsp dried mixed herbs
1 egg
½tsp salt
¼tsp pepper

Tomato sauce:
1tbsp vegetable oil
60g onions, chopped
1 clove garlic, crushed
160g red peppers, sliced
160g yellow peppers, sliced
397g canned tomatoes
4tbsp fresh parsley,
    chopped
½tsp salt
¼tsp pepper

For the sauce, fry the onions and garlic in oil until softened, add peppers and then the remaining ingredients and simmer for 5 minutes. Combine ingredients for the meatballs, divide into 12 and shape into balls. Bake for 10 minutes at 180°C/mark 4, add sauce, cover and cook for a further 30 minutes.

Weight loss: 12%

255 Pork and chicken chow mein

200g pork steaks, strips
200g chicken breast, strips
1tbsp vegetable oil
1 clove garlic, crushed
½ red chilli, finely chopped
75g carrots, sliced
100g mushrooms, sliced
50g prawns, cooked
150g water chestnuts,
   canned, drained and sliced
75g green peppers, sliced
100g fresh beansprouts
400g boiled egg noodles
1tbsp dry sherry
125ml stock
50g spring onions, sliced

Marinade:
½tsp salt
1tsp sugar
4tbsp soy sauce

Marinade pork and chicken for 30 minutes. Stir-fry the garlic, chilli, pork and chicken in oil until browned, add the remaining ingredients and cook for 5-7 minutes.

Weight loss: 11%

256 Pork casserole, made with canned cook-in sauce

675g pork steaks

390g cook-in sauce

Pour the sauce over the pork steaks and cook in a covered casserole dish for 1½ hours at 180°C/mark4.

Weight loss: 20%

257/8 Pork chops in mustard and cream

1tbsp vegetable oil
700g pork loin chops
  (weighed with bone)
50g onions, thinly sliced
250ml dry white wine

125ml stock
2tbsp wholegrain mustard
150ml single cream
½tsp salt
¼tsp pepper

Brown the chops and onions in oil. Add wine and stock, cover and simmer for 15 minutes. Remove pork. Boil remaining stock in the pan for 10-15 minutes, add mustard, cream and seasoning and heat through without boiling. Pour over the pork.

Weight loss: 30% (with bone), 33% (without bone)

259 Pork and pineapple kebabs

400g pork fillet, cubed
125g pineapple, canned
150g green peppers, cubed
100g small cherry tomatoes
100g small button mushrooms

Marinade:
1tbsp vinegar
2tbsp vegetable oil
1tbsp honey
1tsp Worcestershire sauce
½tsp salt
¼tsp pepper

Marinade pork for 2 hours. Thread the pork onto skewers alternately with pineapple, peppers, tomatoes and mushrooms. Grill gently for 30 minutes, turning frequently and brushing with the remaining marinade until cooked.

Weight loss: 35%

260/1 Pork spare ribs, 'barbecue style'

900g pork spare ribs
  (weighed with bone)

Marinade:
150g onions
3 cloves garlic, crushed
3tbsp soy sauce
3tbsp vegetable oil
1tbsp Worcestershire sauce
2tbsp vinegar
½tsp salt
¼tsp pepper

Marinade the pork for 6 hours, turning occasionally. Grill ribs in the marinade for 20-30 minutes under a moderate heat, basting and turning frequently.

Weight loss: 26% (with bone), 43% (without bone)

264/5 Pork spare ribs in black bean sauce

600g pork spare ribs
  (weighed with bone)

Marinade:
20g root ginger, grated
1 clove garlic, crushed
50g black bean sauce
2tsp sugar
2tbsp soy sauce
2tbsp sherry
20ml sesame oil

Marinade the pork for 1 hour. Steam with marinade ingredients for 45 minutes.

Weight loss: 4.9% (with bone), 8.2% (without bone)

266 Pork, stir-fried with vegetables

500g pork fillet, strips
12g root ginger, grated
2 cloves garlic, chopped
1tbsp vegetable oil
100g green peppers, strips
50g fresh beansprouts
60g spring onions, strips
½tsp salt
2tsp dry sherry
40g oyster sauce
1tsp cornflour
3tbsp water

Marinade:
¼tsp salt
½tsp sugar
2tsp soy sauce
¼tsp pepper
1tsp dry sherry
½tsp cornflour
1tbsp water

Marinade the pork for 20 minutes. Stir-fry the ginger, garlic and pork in oil until browned, add green peppers, beansprouts, spring onions, salt, sherry and oyster sauce. Add cornflour dissolved in the water and stir until meat and vegetables are coated.

Weight loss: 17%

267 Rabbit casserole

1tbsp vegetable oil
50g back bacon rashers, chopped
500g rabbit, diced
1tbsp cornflour
600ml stock

150g carrots, sliced
1tbsp fresh parsley,
  chopped
½tsp salt
¼tsp pepper

Brown the bacon and rabbit in oil. Transfer to a casserole dish and add cornflour, stock, carrots, parsley and seasoning. Cover and bake for 1½ hours at 180°C/mark 4.

Weight loss: 21%

268 Salmis of pheasant

500g roast pheasant, meat only

Salmis:
500ml stock
1tsp mixed dried herbs
25g margarine
15g flour
125ml red wine

50g carrots
25g celery
50g onions
160g orange, peeled and
  segmented

Place the salmis ingredients in a pan. Bring to the boil and simmer for 10 minutes. Serve over pheasant.

Weight loss: 6% (sauce only)

269 Sausage casserole

400g diced pork
150g onions, chopped
200g streaky bacon rashers,
   chopped
1tbsp vegetable oil
200g pork sausage, chopped
227g baked beans, in tomato sauce,
   canned

1 bay leaf
1tsp dried mixed herbs
300ml stock
½tsp salt
¼tsp pepper

Brown the pork, onions and bacon in oil, add the remaining ingredients and bake, uncovered, for 1½ hours at 170°C/mark 3.

Weight loss: 15%

270 Shepherd's pie

500g stewed minced lamb
150ml stock
½tsp salt
¼tsp pepper

500g boiled potatoes
50ml milk
20g margarine

Mix the lamb (No 246), stock and seasoning. Cover with potatoes mashed with milk and margarine, and bake for 30 minutes at 190°C/mark 5.

Weight loss: 11%

271 Shish kebab, with onions and peppers

400g lamb neck fillet,
   diced
200g onions, thickly sliced
300g green peppers, diced

Marinade:
1 clove garlic
½tsp salt
1tbsp vinegar
2tbsp vegetable oil
1tbsp garam masala
1 green chilli, chopped

Marinade the lamb for 2 hours. Thread the lamb onto skewers alternately with onions and peppers. Grill for 30 minutes under a moderate grill, turning frequently and brushing with the remaining marinade until cooked.

Weight loss: 35%

272 Spaghetti bolognese

900g bolognese sauce (No 183)

900g boiled spaghetti

Mix ingredients together.

274 Stuffed cabbage leaves

400g minced lamb
150g onions, finely chopped
300g boiled white long grain rice
1 egg
1tbsp fresh mint, chopped

½tsp salt
¼tsp pepper
400g Savoy cabbage leaves
350ml stock

Mix the mince, onions, rice, egg, mint and seasoning together. Fold the leaves around the stuffing and pack the parcels closely together in an ovenproof dish. Pour the stock over, cover and bake for 1 hour at 180°C/mark 4.

Weight loss: 12%

275 Stuffed peppers

300g minced beef
100g onions, finely chopped
300g canned tomatoes
2tsp cornflour
½tsp dried mixed herbs
1tbsp Worcestershire sauce

½tsp salt
¼tsp pepper
150g boiled white long grain rice
500g green peppers, halved
30g dried breadcrumbs

Brown the mince and onions, add the tomatoes, cornflour, herbs, Worcestershire sauce and seasoning. Cover and simmer for 20 minutes. Stir in rice and fill the pepper halves with the mince stuffing, sprinkle with breadcrumbs and bake, uncovered for 35 minutes at 190°C/mark 5.

Weight loss: 14%

276 Sweet and sour pork

400g diced pork

Marinade:
½tsp salt
1tbsp soy sauce
2tbsp sherry
½tsp sugar

Batter:
1tbsp cornflour
1tbsp water
½ egg

(16g vegetable oil)

Sauce:
1tbsp vegetable oil
1 clove garlic, crushed
7g root ginger, grated
100g onions, chopped
75g green peppers, sliced
75g red peppers, sliced
30g sugar
30ml vinegar
2tsp cornflour
1tbsp soy sauce
1tbsp sherry
1tbsp tomato purée
5tbsp water

Marinade the pork for 1 hour. Coat the pork with batter ingredients and deep-fry for 4 minutes. For the sauce, stir-fry garlic, ginger and onions in oil, add the remaining ingredients and cook until thickened. Add pork, stir and heat through.

Weight loss: 28%

277 Sweet and sour pork, made with lean pork

As for sweet and sour pork (No 276), except made with lean diced pork.

278 Sweet and sour pork, made with canned sweet and sour sauce

500g diced pork
1tbsp vegetable oil

390g sweet and sour sauce, canned

Brown the pork in oil. Add sauce and cook for 2 minutes.

Weight loss: 16%

280 Toad in the hole

100g flour
¼tsp salt
2 eggs

200ml milk
400g pork sausages

Mix flour and salt in a bowl and gradually add in the eggs and milk. Prick sausages and bake for 5 minutes at 220°C/mark 7. Spread fat from sausages around dish and pour batter over sausages. Bake for a further 40 minutes.

Weight loss: 21%

281 Toad in the hole with skimmed milk and reduced fat sausages

As for toad in the hole (No 280), except made with skimmed milk and reduced fat sausages.

Weight loss: 21%

282 Tripe and onions, stewed

450g dressed tripe, washed,
  chopped
200g onions, sliced
300ml milk

½tsp salt
¼tsp pepper
30g flour

Simmer the tripe, onions, milk and seasoning for about 2 hours, until tender, stir in the flour and cook until thickened.

Weight loss: 43%

283 Turkey and pasta bake

1tbsp vegetable oil
100g back bacon rashers,
  chopped
60g onions, finely chopped
400g cooked light and dark turkey
  meat, diced
200g mushrooms, sliced
160g red peppers, chopped
1tsp dried mixed herbs
½tsp salt
¼tsp pepper

Sauce:
15g flour
15g butter
400ml milk

400g boiled macaroni

To top:
75g cheese, grated

Brown the onions and bacon in oil, add turkey, mushrooms, peppers, dried mixed herbs and seasoning and cook for 2 minutes. For the white sauce, melt the butter and cook with the flour for 1 minute, add milk and stir until thickened. Add macaroni and meat mixture, stir and transfer to an ovenproof dish. Top with grated cheese and bake for 25 minutes at 200°C/mark 6.

Weight loss: 10%

284 Turkey, stir-fried with vegetables

1tbsp vegetable oil
350g light and dark turkey meat,
  strips
100g broccoli florets
100g baby sweetcorn
40g spring onions, sliced
150g red peppers, sliced
100g fresh beansprouts

Sauce:
100ml stock
1tbsp dry sherry
1tbsp light soy sauce
5g tomato purée
2tsp cornflour

Stir-fry the turkey in oil until browned, add broccoli, sweetcorn, spring onions, peppers and beansprouts. Add the sauce ingredients and simmer for 3 minutes.

Weight loss: 17%

285 Venison in red wine and port

500g venison, diced
200ml red wine
100g onions, sliced
100g carrots, sliced
25g butter
25g flour

400ml stock
40ml port
2 bay leaves
½tsp salt
¼tsp pepper

Marinade the venison in red wine for 8 hours. Brown the onions and carrots in butter, add the drained venison and cook for 10 minutes, turning once. Transfer to a casserole dish. Mix the flour, stock, port, remaining marinade, bay leaves and seasoning together, bring to the boil. Pour over the venison, cover and cook for 1 hour at 170°C/mark 3.

Weight loss: 14%

286 Wiener schnitzel

300g veal escalopes
15g flour
¼tsp salt
¼tsp pepper

40g egg
50g dried breadcrumbs
2tbsp vegetable oil

Roll escalopes in seasoned flour, dip in egg and coat with the breadcrumbs. Fry in oil for about 10 minutes, turning once.

Weight loss: 12%

INGREDIENT CODES OF FOODS USED IN RECIPES

The ingredients used in the recipes are listed below in alphabetical order.

The code numbers on the left hand side have been taken from the earlier food group supplements and Fifth Edition of 'The Composition of Foods'. Only one code has been given per food, representing the most recent occurrence of a food in the publication series. The codes will therefore direct users to the most relevant data sets used in the recipe calculations. Codes include the 2-digit prefix appropriate to that supplement.

The prefixes for each publication are:

Cereals and Cereal Products	11
Milk Products and Eggs	12
Vegetables, Herbs and Spices	13
Fifth Edition	50
Fruit and Nuts	14
Fish and Fish Products	16
Miscellaneous Foods	17
Meat, Poultry and Game	18

Code number	Ingredient name	Code number	Ingredient name
14-801	Almonds, ground	50-718	Beans, red kidney, canned
14-271	Apple juice	50-696	Beansprouts, mung, raw
14-012	Apples, eating, flesh and skin, raw	17-207	Beer (Bitter)
11-001	Arrowroot	17-291	Black bean sauce
50-739	Aubergine, raw	17-247	Brandy
19-002	Bacon rashers, back, dry-fried	50-048	Bread, white, average (fresh crumbs)
19-001	Bacon rashers, back, raw	50-056	Bread, wholemeal, average
19-014	Bacon rashers, middle, fried	11-069	Breadcrumbs, packet
19-016	Bacon rashers, streaky, raw	50-744	Broccoli, green, raw
50-694	Baked beans, canned in tomato sauce	50-061	Buns, hamburger
17-355	Baking powder	17-013	Butter
13-163	Bamboo shoots, canned, drained	13-192	Cabbage, Savoy, raw
18-006	Beef, braising steak, raw, lean	50-754	Carrots, raw
18-007	Beef, braising steak, raw, lean and fat	50-755	Carrots, boiled
18-017	Beef, fillet steak, raw, lean and fat	14-812	Cashew nuts, roasted
18-040	Beef, mince, extra lean, raw	50-761	Celery, raw
18-036	Beef, mince, raw	50-228	Cheese, Cheddar, average
18-044	Beef, rump steak, raw, lean and fat	50-247	Cheese, Parmesan
18-076	Beef, stewing steak, raw, lean		Chicken breast, raw see Chicken, light meat, raw
18-077	Beef, stewing steak, raw, lean and fat		
18-085	Beef, topside, raw, lean and fat	18-293	Chicken, leg quarter, raw, meat and skin

Code number	Ingredient name
18-294	Chicken, leg quarter, raw, meat and skin, weighed with bone
18-290	Chicken, light meat, raw
18-330	Chicken, light meat, roasted
18-331	Chicken, light and dark meat, average, roasted
17-222	Cider, dry
14-818	Coconut, desiccated
17-295	Cook-in-sauces, canned
19-128	Corned beef, canned
50-004	Cornflour
50-212	Cream, single
50-213	Cream, soured
17-297	Curry paste
17-298	Curry sauce, canned
18-374	Duck, roasted, meat, fat and skin
18-372	Duck, roasted, meat only
50-290	Egg, raw
50-291	Egg white, raw
50-028	Egg noodles, boiled
50-015	Flour, white, self-raising
50-014	Flour, white, plain
19-100	Frankfurter
50-772	Garlic, raw
13-247	Ginger root
18-396	Heart, lamb, raw
17-505	Honey
	Hotdog buns *see* Buns, Hamburger
17-075	Jam, reduced sugar
17-074	Jam, stone fruit
18-402	Kidney, lamb, raw
18-158	Lamb mince, raw
18-160	Lamb, neck fillet, raw, lean
18-161	Lamb, neck fillet, raw, lean and fat
18-182	Lamb, stewing, raw, lean and fat
17-010	Lard
11-051	Lasagne, raw
14-277	Lemon juice
14-127	Lemon peel
13-090	Lentils, green and brown, whole, boiled in unsalted water
18-412	Liver, chicken, fried
18-413	Liver, lamb, raw
18-414	Liver, lamb, fried
50-026	Macaroni, boiled
17-318	Mayonnaise, reduced calorie

Code number	Ingredient name
17-316	Mayonnaise, retail
50-185	Milk, semi-skimmed
50-181	Milk, skimmed
50-189	Milk, whole
50-783	Mushrooms, raw
17-364	Mustard, smooth
17-365	Mustard, wholegrain
17-363	Mustard powder, made up
50-792	Onion, raw
50-795	Onions, fried in corn oil
14-283	Orange juice, unsweetened
17-321	Oyster sauce
50-846	Parsley, fresh
17-323	Pasta sauce, tomato based
17-322	Pasta sauce, white, with ham and mushrooms
19-143	Pâté, liver
50-730	Peas, boiled
13-132	Peas, frozen, raw
50-801	Pepper, chilli, green, raw
13-317	Pepper, chilli, red, raw
50-802	Pepper, green, raw
50-804	Pepper, red, raw
13-322	Pepper, yellow, raw
18-383	Pheasant, roasted, meat only
14-211	Pineapple, canned in juice
18-217	Pork, diced, raw, lean
18-218	Pork, diced, raw, lean and fat
18-225	Pork, fillet, raw
18-246	Pork, loin chops, raw, lean and fat
18-247	Pork, loin chops, raw, lean and fat, weighed with bone
18-267	Pork, mince, raw
18-278	Pork, spare ribs, raw, lean and fat
18-279	Pork, spare ribs, raw, lean and fat, weighed with bone
18-269	Pork, spare rib chops, raw, lean and fat
18-270	Pork, spare rib chops, raw, lean and fat, weighed with bone
18-277	Pork, spare rib joint, pot-roasted, lean and fat
18-283	Pork, steaks, raw, lean only
17-234	Port
50-664	Potatoes, raw
50-668	Potatoes, boiled in unsalted water
16-238	Prawns

Code number	Ingredient name
18-387	Rabbit, raw, meat only
18-388	Rabbit, stewed, meat only
50-718	Red kidney beans, canned, drained
17-325	Redcurrant jelly
11-046	Rice, white, glutinous, raw
50-022	Rice, white, long grain, raw
50-023	Rice, white, long grain, boiled
50-1184	Salt
19-078	Sausages, pork, raw
19-084	Sausages, pork, reduced fat, raw
17-043	Sesame oil
13-342	Shallots, raw
17-235	Sherry, dry
17-236	Sherry, medium
50-1171	Soy sauce
50-030	Spaghetti, white, boiled
50-813	Spinach, raw
50-818	Spring onions, raw
17-011	Suet, shredded
17-060	Sugar, brown
17-063	Sugar, white
50-819	Swede, raw
17-335	Sweet and sour sauce, canned

Code number	Ingredient name
50-823	Sweetcorn, baby, canned, drained
13-388	Tomatoes, cherry, raw
50-827	Tomatoes, raw
50-832	Tomatoes, canned
17-374	Tomato purée
11-093	Tortillas
18-428	Tripe, dressed, raw
18-350	Turkey, meat, average, raw
18-361	Turkey, meat, average, roasted
13-861	Turmeric, ground
50-833	Turnip, raw
18-092	Veal escalopes, raw
17-046	Vegetable oil, blended average
18-390	Venison, raw
18-391	Venison, roasted
17-339	Vinegar
17-340	Worcestershire sauce
17-377	Water
13-395	Water chestnuts, canned, drained
17-228	Wine, red
17-231	Wine, white
50-255	Yogurt, plain, low fat
50-260	Yogurt, whole milk, plain

INDIVIDUAL FATTY ACIDS

The amounts of the main fatty acids in typical meat products and dishes are given here in grams per 100g food in contrast to those for Meat, Poultry and Game (Chan *et al*, 1995) which were given in grams per 100g fatty acids. Values have been obtained by analysis only, and do not include values derived by calculation or estimation. The values for the unsaturated fatty acids include both *cis* and *trans* fatty acids and the various positional isomers.

Fatty acid conversion factors for meat products and dishes in the main tables have been derived from those given below:

Conversion factors to give total fatty acids in fat

Beef lean	0.916	Pork lean	0.910
Beef fat	0.953	Pork fat	0.953
Lamb lean	0.916	Poultry	0.945
Lamb fat	0.953	Fats and oils	0.956

Names of fatty acids occurring in the tables:

No of carbon atoms and double bonds	Systematic name	Common name
Saturated acids		
4:0	Butanoic acid	Butyric acid
6:0	Hexanoic acid	Caproic acid
10:0	Decanoic acid	Capric acid
12:0	Dodecanoic acid	Lauric acid
14:0	Tetradecanoic acid	Myristic acid
15:0	Pentadecanoic acid	
16:0	Hexadecanoic acid	Palmitic acid
17:0	Heptadecanoic acid	Margaric acid
18:0	Octadecanoic acid	Stearic acid
20:0	Eicosanoic acid	Arachidic acid Arachic acid
22:0	Docosanoic acid	Behenic acid

No of carbon atoms and double bonds	Systematic name	Common name
Monounsaturated acids		
14:1	Tetradecenoic acid	Myristoleic acid
16:1	Hexadecenoic acid	Palmitoleic acid
17:1	Heptadecenoic acid	
18:1 (*cis*)	Octadecenoic acid	Oleic acid
		cis-Vaccenic acid
18:1 (*trans*)		Elaidic acid
		trans-Vaccenic acid
20:1	Eicosenoic acid	Eicosenic acid
		Gadoleic acid
22:1	Docosenoic acid	Erucic acid
Polyunsaturated acids		
18:2	Octadecadienoic acid	Linoleic acid
18:3	Octadecatrienoic acid	Linolenic acid
18:4	Octadecatetraenoic acid	Stearidonic acid
20:2	Eicosadienoic acid	
20:3	Eicosatrienoic acid	
20:4	Eicosatetraenoic acid	Arachidonic acid
20:5	Eicosapentaenoic acids	
22:2	Docosadienoic acid	
22:5	Docosapentaenoic acid	Clupanodonic acid
22:6	Docosahexaenoic acid	Cervonic acid

Meat products and dishes

Fatty acids, g per 100g food

No. Food 19-	Saturated					Monounsaturated				
	14:0	15:0	16:0	17:0	18:0	14:1	16:1	17:1	18:1	20:1
Bacon and ham										
1 **Bacon rashers, back**, *raw*	0.2	Tr	3.8	Tr	2.1	Tr	0.4	Tr	6.4	0.1
3 -, *grilled*	0.3	Tr	4.9	0.1	2.8	Tr	0.5	0.1	8.3	0.2
13 **middle**, *raw*	0.3	Tr	4.4	0.1	2.4	Tr	0.5	0.1	7.9	0.2
14 -, *fried* [a]	0.4	Tr	6.0	0.1	3.2	Tr	0.7	0.1	10.9	0.3
15 -, *grilled*	0.3	Tr	5.0	0.1	2.8	Tr	0.6	0.1	9.1	0.2
16 **streaky**, *raw*	0.3	Tr	5.1	0.1	2.5	Tr	0.7	0.1	9.3	0.2
18 -, *grilled*	0.4	Tr	6.1	0.1	3.1	Tr	0.7	0.1	10.4	0.2
19 **Bacon loin steaks**, *grilled*	0.1	Tr	2.2	Tr	1.1	Tr	0.2	Tr	3.7	0.1
21 **Ham, gammon joint**, *boiled*	0.2	Tr	2.6	Tr	1.3	Tr	0.4	Tr	4.9	0.1
22 -, **gammon rashers**, *grilled*	0.1	Tr	2.1	Tr	1.2	Tr	0.2	Tr	3.7	0.1
23 **Ham**	Tr	Tr	0.7	Tr	0.4	Tr	0.1	Tr	1.3	Tr
26 -, *premium*	0.1	Tr	1.1	Tr	0.6	Tr	0.1	Tr	2.0	Tr
27 **Pork shoulder**, cured, slices	Tr	Tr	0.8	Tr	0.4	Tr	0.1	Tr	1.4	Tr
Burgers and grillsteaks										
28 **Beefburgers**, chilled/frozen, *raw*	0.9	0.1	6.3	0.3	3.5	0.2	1.1	0.2	9.7	0.1
29 -, -, *fried*	0.7	0.2	5.8	0.5	3.4	0.2	1.1	0.2	9.3	0.1
35 **low fat**, chilled/frozen, *raw*	0.3	0.1	2.3	0.2	1.4	0.1	0.4	0.1	3.6	Tr
36 -, -, *fried*	0.3	0.1	2.7	0.2	1.6	0.1	0.4	0.1	4.0	Tr

[a] Contains 0.1g 20:0 per 100g food

Fatty acids, g per 100g food

No. 19-	Food	Polyunsaturated				
		18:2	18:3	20:2	20:3	20:4
	Bacon and ham					
1	**Bacon rashers, back**, *raw*	1.9	0.2	0.1	Tr	Tr
3	-, *grilled*	2.3	0.2	0.1	0.1	0.1
13	**middle**, *raw*	2.1	0.2	0.1	Tr	Tr
14	-, *fried* [a]	4.1	0.3	0.1	0.1	0.1
15	-, *grilled*	2.4	0.3	0.1	0.1	0.1
16	**streaky**, *raw*	2.9	0.3	0.1	0.1	0.1
18	-, *grilled* [b]	3.0	0.3	0.1	0.1	0.1
19	**Bacon loin steaks**, *grilled*	1.1	0.1	Tr	Tr	0.1
21	**Ham, gammon joint**, *boiled*	1.5	0.2	0.1	Tr	0.1
22	**-, gammon rashers**, *grilled*	1.3	0.1	0.1	Tr	0.1
23	**Ham**	0.4	Tr	Tr	Tr	Tr
26	-, premium	0.6	0.1	Tr	Tr	0.1
27	**Pork shoulder**, cured, slices	0.5	Tr	Tr	Tr	Tr
	Burgers and grillsteaks					
28	**Beefburgers**, chilled/frozen, *raw*	0.5	0.1	0	0	Tr
29	-, *fried*	0.6	0.2	0	0	0
35	low fat, chilled/frozen, *raw*	0.3	0.1	0	0	Tr
36	-, *fried*	0.3	0.1	0	0	Tr

[a] Contains 0.1g 22:5, 0.1g 22:6 per 100g food

[b] Contains 0.1g 22:2 per 100g food

Meat products and dishes

Fatty acids, g per 100g food

No. Food 1g-	Saturated					Monounsaturated				
	14:0	15:0	16:0	17:0	18:0	14:1	16:1	17:1	18:1	20:1
Burgers and grillsteaks continued										
38 **Beefburgers in gravy**, canned	0.2	0.1	2.5	0.2	1.7	0.1	0.4	0.1	4.6	0.1
43 **Economy burgers**, frozen, *grilled*	0.4	0.1	4.4	0.2	2.1	0.1	0.8	0.1	7.5	0.1
44 **Grillsteaks, beef**, chilled/frozen, *raw*	0.7	0.2	5.8	0.5	3.4	0.2	1.0	0.2	9.3	Tr
49 **Steaklets**, frozen, *raw*	0.6	0.2	5.6	0.5	3.2	0.2	1.2	0.2	9.8	0.1
Meat pies and pastries										
51 **Beef pie**, chilled/frozen, *baked* [a,b]	0.6	0.1	4.4	0.2	2.7	0.1	0.7	0.1	7.1	0.6
53 **Bridie/Scotch pie**, individual	0.3	0.1	2.5	0.2	1.6	0.1	0.4	0.1	3.9	0.1
55 **Chicken pie**, individual, chilled/frozen, *baked* [c,d]	0.3	Tr	4.9	0.1	1.5	Tr	0.3	Tr	6.8	0.2
56 **Cornish pastie** [e]	0.5	0.1	3.0	0.2	1.8	Tr	0.6	0.1	7.1	0.6
62 **Pork and egg pie** [c,d]	0.3	0	4.3	0.1	2.5	0	0.5	0.1	8.2	0.3
63 **Pork pie**, individual [f]	0.5	Tr	5.9	0.1	3.0	Tr	0.7	0.1	9.9	0.2
64 -, mini [g]	0.6	0.1	6.6	0.2	3.6	Tr	0.9	0.1	10.2	0.3
65 -, sliced [f]	0.4	Tr	6.7	0.1	4.0	0	0.8	0.1	12.1	0.3
69 **Steak and kidney/Beef pie**, individual, chilled/frozen, *baked* [c,d]	0.4	0.1	5.6	0.1	2.0	Tr	0.4	0.1	7.0	0.2
72 **Steak and kidney pudding**, canned [d,f]	0.4	0.2	2.4	0.4	1.8	0.1	0.4	0.1	3.4	0.3

[a] Contains 0.3g 20:0, 0.2g 22:0 per 100g food
[c] Contains 0.1g 20:0, 0.1g 22:0 per 100g food
[e] Contains 0.2g 20:0 per 100g food
[g] Contains 0.1g 12:0, 0.1g 20:0 per 100g food

[b] Contains 0.4g 22:1 per 100g food
[d] Contains 0.1g 22:1 per 100g food
[f] Contains 0.1g 20:0 per 100g food

Meat products and dishes

Fatty acids, g per 100g food

No. 19-	Food	Polyunsaturated 18:2	18:3	20:2	20:3	20:4
	Burgers and grillsteaks continued					
38	**Beefburgers in gravy**, canned	0.5	0.1	0	0	Tr
43	**Economy burgers**, frozen, *grilled*	1.6	0.2	Tr	Tr	Tr
44	**Grillsteaks, beef**, chilled/frozen, *raw*	0.7	0.2	0	0	0
49	**Steaklets**, frozen, *raw*	0.5	0.1	0	Tr	0.1
	Meat pies and pastries					
51	**Beef pie**, chilled/frozen, *baked*	1.7	0.2	0.1	Tr	Tr
53	**Bridie/Scotch pie**, individual	0.2	0.1	0	Tr	Tr
55	**Chicken pie**, individual, chilled/frozen, *baked*	2.1	0.2	Tr	Tr	Tr
56	**Cornish pastie**	1.1	0.2	0	0	0
62	**Pork and egg pie**	2.0	0.2	0.1	0	0.1
63	**Pork pie**, individual [a]	2.8	Tr	0.1	Tr	0.1
64	-, mini	3.0	0.3	0.1	0.1	0.1
65	-, sliced	2.9	0.3	0.1	0	0
69	**Steak and kidney/Beef pie**, individual, chilled/frozen, *baked*	1.6	0.2	Tr	Tr	Tr
72	**Steak and kidney pudding**, canned [b]	0.5	0.2	0.1	0	Tr

[a] Contains 0.2g 18:4 per 100g food

[b] Contains 2.3g unidentified fatty acids per 100g food

145

Meat products and dishes

Fatty acids, g per 100g food

No. Food	Saturated					Monounsaturated				
19-	14:0	15:0	16:0	17:0	18:0	14:1	16:1	17:1	18:1	20:1
Sausages										
75 **Beef sausages**, chilled, *raw*	0.5	0.2	5.5	0.3	2.9	0.2	0.9	0.2	9.6	0.1
76 -, *fried*	0.4	0.1	4.4	0.3	2.2	0.1	0.8	0.1	8.0	0.1
77 -, *grilled*	0.5	0.1	4.6	0.3	2.4	0.1	0.8	0.1	7.7	0.1
78 **Pork sausages**, chilled, *raw*	0.4	Tr	5.3	0.1	2.7	Tr	0.6	0.1	9.3	0.2
79 -, *fried*	0.3	Tr	5.2	0.1	2.7	Tr	0.6	0.1	9.4	0.2
81 frozen, *raw* [a]	0.4	Tr	6.1	0.1	3.1	Tr	0.7	0.1	10.9	0.3
84 reduced fat, chilled/frozen, *raw*	0.1	Tr	2.3	0	1.2	Tr	0.3	Tr	4.1	0.1
85 -, -, *fried*	0.2	Tr	2.6	0.1	1.3	Tr	0.3	Tr	5.2	0.1
87 **Pork and beef sausages**, chilled, *raw*	0.3	0.1	5.0	0.2	2.8	Tr	0.7	0.1	9.3	0.2
88 -, *grilled* [a]	0.3	Tr	4.4	0.1	2.4	Tr	0.6	0.1	8.2	0.2
89 frozen, *raw* [a]	0.5	0.1	6.8	0.2	3.7	Tr	0.9	0.1	11.5	0.2
90 -, **economy**, chilled, *raw*	0.3	Tr	4.6	0.2	2.4	0	0.6	0	8.3	0.2
94 **Premium sausages**, chilled, *fried*	0.3	Tr	4.6	0.2	2.5	Tr	0.5	0	7.9	0.1
Continental style sausages										
100 **Frankfurter**	0.4	Tr	5.6	0.1	3.0	Tr	0.7	0.1	10.5	0.2
103 **Garlic sausage**	0.3	Tr	4.5	0.1	2.3	0	0.6	0.1	8.3	0.2
106 **Liver sausage** [a]	0.2	0.1	2.9	0.1	1.9	Tr	0.4	Tr	4.9	0.4
108 **Pepperami** [a]	0.8	Tr	11.7	0.4	6.4	0.1	1.4	0.1	21.1	0.4

[a] Contains 0.1g 20:0 per 100g food

No. Food	Polyunsaturated				
19-	18:2	18:3	20:2	20:3	20:4
Sausages					
75 **Beef sausages**, chilled, *raw*	1.6	0.2	Tr	0	Tr
76 -, *fried*	1.5	0.2	Tr	Tr	Tr
77 -, *grilled*	1.2	0.1	Tr	0	0
78 **Pork sausages**, chilled, *raw*	2.8	0.3	0.1	Tr	Tr
79 -, *fried* [a]	2.9	Tr	0.1	Tr	Tr
81 frozen, *raw*	3.1	0.3	0.1	0.1	0.1
84 reduced fat, chilled/frozen, *raw*	1.4	0.1	0	Tr	Tr
85 -, -, *fried*	2.0	0.2	Tr	0	Tr
87 **Pork and beef sausages**, chilled, *raw*	2.1	0.2	0.1	Tr	Tr
88 -, *grilled*	1.9	0.2	0.1	Tr	0.1
89 frozen, *raw*	2.4	0.2	0	0	0.1
90 -, **economy**, chilled, *raw*	2.3	0.2	0.1	0	Tr
94 **Premium sausages**, chilled, *fried*	2.6	0.2	0.1	Tr	0.1
Continental style sausages					
100 **Frankfurter**	2.5	0.2	0.1	Tr	0.1
103 **Garlic sausage**	2.1	0.2	0.1	Tr	0.1
106 **Liver sausage** [b]	1.5	0.2	0.1	0.1	0.2
108 **Pepperami**	4.4	0.4	0.2	0.1	0

[a] Contains 0.3g 18:4 per 100g food

[b] Contains 0.1g 20:5, 0.1g 22:5, 0.1g 22:6 per 100g food

Meat products and dishes

No. 19-	Food	Saturated					Monounsaturated				
		14:0	15:0	16:0	17:0	18:0	14:1	16:1	17:1	18:1	20:1
	Continental style sausages *continued*										
110	**Salami** [a,b]	0.5	Tr	8.8	0.2	4.9	Tr	1.0	0.1	16.1	0.4
111	**Saveloy**, unbattered, takeaway [b]	0.3	0	4.6	0.1	2.4	0	0.6	0.1	8.7	0.2
	Other meat products										
113	**Black pudding**, *raw*	0.3	Tr	4.8	0.1	2.5	Tr	0.6	0.1	8.6	0.2
116	**Chicken in crumbs**, stuffed with cheese and vegetables, chilled/frozen, *baked* [c]	0.3	Tr	2.5	Tr	0.7	Tr	0.3	Tr	4.1	Tr
117	**Chicken breast in crumbs**, chilled, *raw*	Tr	Tr	1.3	Tr	0.4	0	Tr	Tr	3.0	Tr
118	-, *fried*	Tr	0	1.5	Tr	0.4	0	0.1	0	5.1	0.1
120	**Chicken breast marinated with garlic and herbs**, chilled/frozen, *baked* [d]	0.2	Tr	1.8	0.1	0.6	Tr	0.3	Tr	3.5	Tr
121	**Chicken fingers**, baked	0.1	Tr	2.0	Tr	0.8	Tr	0.4	Tr	5.0	0.1
123	**Chicken kiev**, frozen, *baked* [e]	0.8	0.1	3.6	0.1	1.5	Tr	0.3	Tr	5.0	0.1
124	**Chicken nuggets**, takeaway	0.1	Tr	1.8	Tr	1.2	Tr	0.2	Tr	6.5	0.1
125	**Chicken roll**	0.1	Tr	1.1	Tr	0.3	Tr	0.2	0	1.8	0.1
127	**Chicken tandoori**, chilled, *reheated*	0.1	Tr	2.4	Tr	0.7	Tr	0.5	Tr	4.3	0.1
128	**Corned beef**, canned [f]	0.4	0.1	3.2	0.3	1.7	0.1	0.4	0.1	3.7	Tr
129	**Doner kebab**, meat only [g]	1.4	0.4	6.4	0.8	6.0	Tr	0.7	0.2	10.9	0.1
131	**Faggots in gravy**, chilled/frozen, *reheated*	0.1	Tr	1.5	Tr	0.8	Tr	0.2	Tr	2.7	0.1

[a] Contains 0.1g 22:0 per 100g food
[b] Contains 0.1g 22:1 per 100g food
[c] Contains 0.1g 6:0, 0.1g 10:0, 0.1g 12:0 per 100g food
[d] Contains 0.1g 12:0 per 100g food
[e] Contains 0.1g 4:0, 0.1g 6:0, 0.2g 10:0, 0.2g 12:0 per 100g food
[f] Contains 0.1g 20:0 per100g food
[g] Contains 0.1g 10:0, 0.1g 12:0, 0.1g 20:0 per 100g food

No. Food	Polyunsaturated					
19-	18:2	18:3	20:2	20:3	20:4	
Continental style sausages continued						
110 **Salami**	3.9	0.4	0.1	Tr	Tr	
111 **Saveloy,** unbattered, takeaway	3.2	0.4	0	0	0	
Other meat products						
113 **Black pudding,** raw	2.1	0.2	0.1	Tr	Tr	
116 **Chicken in crumbs,** stuffed with cheese and						
vegetables, chilled/frozen, *baked*	3.8	0.4	Tr	Tr	Tr	
117 **Chicken breast in crumbs,** chilled, *raw*	2.8	0.3	0	0	Tr	
118 -, *fried*	4.0	0.6	0	0	0	
120 **Chicken breast marinated with garlic and**						
herbs, chilled/frozen, *baked*	1.4	0.2	Tr	Tr	Tr	
121 **Chicken fingers,** baked	3.1	0.2	0	0	0	
123 **Chicken kiev,** frozen, *baked*	3.1	0.4	Tr	Tr	Tr	
124 **Chicken nuggets,** takeaway	2.0	0.2	0	Tr	Tr	
125 **Chicken roll**	0.7	0.1	0	0	0.1	
127 **Chicken tandoori,** chilled, *reheated*	1.7	0.2	Tr	Tr	Tr	
128 **Corned beef,** canned	0.3	Tr	0	0	0	
129 **Doner kebab,** meat only [a]	0.7	0.5	0	0	Tr	
131 **Faggots in gravy,** chilled/frozen, *reheated*	0.8	0.1	Tr	Tr	Tr	

[a] Contains 0.1g 22:5 per 100g food

Meat products and dishes

Fatty acids, g per 100g food

No. 19-	Food	Saturated					Monounsaturated				
		14:0	15:0	16:0	17:0	18:0	14:1	16:1	17:1	18:1	20:1
	Other meat products continued										
133	**Ham and pork**, chopped, canned	0.3	Tr	5.2	Tr	2.6	0	0.7	Tr	9.5	0.2
135	**Luncheon meat**, canned [a]	0.3	0	4.9	0.2	2.9	0	0.7	0.2	9.7	0.4
138	**Meat loaf**, chilled/frozen, *reheated* [b]	0.2	Tr	3.2	0.1	1.8	0	0.4	0.1	6.3	0.3
139	**Meat spread**	0.3	0.1	3.1	0.2	1.9	Tr	0.5	0.1	5.2	0.1
140	**Mince in gravy**, canned	0.3	0.1	2.6	0.2	1.7	0.1	0.5	0.1	4.6	Tr
141	**Minced beef pie filling**, canned	0.2	0.1	1.9	0.2	1.6	Tr	0.3	0.1	3.2	Tr
143	**Pâté, liver** [c]	0.4	Tr	5.7	0.1	3.2	0	0.7	0.1	10.8	0.2
144	**-, in a tube** [d]	0.4	Tr	4.5	0.1	2.3	0	0.6	0.1	7.8	0.3
145	**-, meat**, low fat	0.1	Tr	2.1	Tr	1.2	Tr	0.2	Tr	3.6	0.1
146	**Pork haslet**	0.2	Tr	2.7	Tr	1.4	Tr	0.3	Tr	4.8	0.1
149	**Rissoles**, savoury	0.5	0.2	3.7	0.4	2.6	0.1	0.6	0.1	6.4	0.1
150	**Shish kebab**, meat only	0.3	0.1	1.9	0.2	1.4	Tr	0.2	0.1	4.0	Tr
152	**Stewed steak with gravy**, canned	0.3	0.1	2.4	0.2	1.6	0.1	0.4	0.1	3.7	0.1
154	**Tongue slices**	0.4	0.2	3.0	0.4	2.1	0.1	0.4	0.2	5.7	Tr
155	**Turkey roast**, frozen, *cooked*	0.1	Tr	1.8	Tr	0.6	Tr	0.5	Tr	2.5	Tr
156	**Turkey roll** [b]	0.1	Tr	1.8	Tr	0.7	Tr	0.4	Tr	3.2	0.1
158	**Turkey steaks in crumbs**, frozen, *grilled*	0.1	Tr	3.2	Tr	0.8	Tr	0.3	Tr	6.4	0.1

[a] Contains 0.3g 20:0 per 100g food
[c] Contains 0.1g 12:0 per 100g food
[b] Contains 0.1g 20:0 per 100g food
[d] Contains 0.1g 12:0, 0.1g 20:0 per 100g food

Fatty acids, g per 100g food

No. Food	Polyunsaturated				
19-	18:2	18:3	20:2	20:3	20:4

Other meat products continued

No. Food	18:2	18:3	20:2	20:3	20:4
133 **Ham and pork**, chopped, canned	2.0	0.1	0	0	0.1
135 **Luncheon meat**, canned	2.4	0.4	0.1	0	0.1
138 **Meat loaf**, chilled/frozen, *reheated* [a]	1.5	0.1	0.1	Tr	0.1
139 **Meat spread**	1.0	0.1	Tr	Tr	Tr
140 **Mince in gravy**, canned	0.4	0.1	0	0	0
141 **Minced beef pie filling**, canned	0.2	0.1	0	0	0
143 **Pâté, liver**	2.4	0.2	0.1	0.2	0
144 -, in a tube [b]	2.0	0.3	0.1	0	0.2
145 -, **meat**, low fat	1.2	0.1	Tr	Tr	0.1
146 **Pork haslet**	1.7	0.2	0.1	Tr	Tr
149 **Rissoles**, savoury	0.5	0.1	0	0	0
150 **Shish kebab**, meat only	0.4	0.2	0	0	Tr
152 **Stewed steak with gravy**, canned	0.2	0.1	0	0	0
154 **Tongue slices**	0.6	0.1	Tr	Tr	0.1
155 **Turkey roast**, frozen, *cooked* [c]	0.9	Tr	0	0	Tr
156 **Turkey roll**	1.6	0.2	Tr	Tr	0.1
158 **Turkey steaks in crumbs**, frozen, *grilled*	4.3	0.7	Tr	Tr	Tr

[a] Contains 0.1g 22:5 per 100g food
[c] Contains 0.1g 18:4 per 100g food

[b] Contains 0.1g 22:5, 0.1g, 22:6 per 100g food

Meat products and dishes

No. 19-	Food	Saturated					Monounsaturated				
		14:0	15:0	16:0	17:0	18:0	14:1	16:1	17:1	18:1	20:1
Meat dishes											
184	**Cannelloni**, chilled/frozen, *reheated* [a]	0.2	Tr	1.0	0.1	0.6	Tr	0.1	Tr	1.8	Tr
194	**Chicken in white sauce**, canned	0.1	Tr	1.7	Tr	0.5	Tr	0.4	Tr	3.4	Tr
216	**Cottage/Shepherd's pie**, chilled/frozen, *reheated*	0.2	0.1	1.2	0.1	0.7	Tr	0.2	Tr	1.9	Tr
224	**Irish stew**, canned	0.1	0.1	1.1	0.1	1.0	Tr	0.1	Tr	1.8	Tr
231	**Lamb/Beef hot pot with potatoes**, chilled/frozen, retail, *reheated*	0.1	Tr	0.9	0.1	0.6	Tr	0.1	Tr	1.8	Tr
238	**Lasagne**, chilled/frozen, *reheated* [b]	0.3	0.1	1.3	0.1	0.6	Tr	0.1	Tr	2.0	Tr
248	**Moussaka**, chilled/frozen/longlife, *reheated* [c]	0.3	0.1	1.3	0.1	0.8	Tr	0.1	Tr	3.4	Tr
249	**Pancakes, beef**, frozen, *shallow-fried*	0.1	Tr	1.5	Tr	0.5	Tr	0.2	Tr	6.2	0.1
250	-, **chicken**, frozen, *shallow-fried*	Tr	Tr	1.0	Tr	0.3	Tr	0.1	0	5.4	0.1
262	**Pork spare ribs, 'barbecue style'**, chilled/frozen, *reheated*	0.2	Tr	3.7	0.1	2.2	Tr	0.4	0	6.2	0.1
273	**Spaghetti bolognese**, chilled/frozen, *reheated*	0.1	0.1	1.2	0.1	0.8	Tr	0.2	Tr	2.2	Tr
279	**Tagliatelle with ham, mushroom and cheese**, chilled/frozen/longlife, *reheated* [d]	0.4	0.1	1.3	0.1	0.5	Tr	0.1	Tr	1.4	Tr

[a] Contains 0.1g 12:0 per 100g food
[b] Contains 0.1g 4:0, 0.1g 10:0, 0.1g 12:0 per 100g food
[c] Contains 0.1g 10:0, 0.1g 12:0 per 100g food
[d] Contains 0.1g 4:0, 0.1g 6:0, 0.1g 10:0, 0.1g 12:0 per 100g food

No. 19-	Food	Polyunsaturated				
		18:2	18:3	20:2	20:3	20:4
	Meat dishes					
184	**Cannelloni**, chilled/frozen, *reheated*	0.5	0.1	0	0	0
194	**Chicken in white sauce**, canned	1.4	0.1	0	Tr	0.1
216	**Cottage/Shepherd's pie**, chilled/frozen, *reheated*	0.3	0.1	0	Tr	Tr
224	**Irish stew**, canned	0.2	0.1	0	0	Tr
231	**Lamb/Beef hot pot with potatoes**, chilled/frozen, retail, *reheated*	0.4	0.1	0	Tr	Tr
238	**Lasagne**, chilled/frozen, *reheated*	0.6	0.1	Tr	Tr	Tr
248	**Moussaka**, chilled/frozen/longlife, *reheated*	0.8	0.3	Tr	0	Tr
249	**Pancakes, beef**, frozen, *shallow-fried*	4.1	0.8	0	Tr	Tr
250	-, **chicken**, frozen, *shallow-fried*	3.6	0.8	Tr	0	Tr
262	**Pork spare ribs, 'barbecue style'**, chilled/frozen, *reheated* [a]	2.2	0.2	0.1	0	0.1
273	**Spaghetti bolognese**, chilled/frozen, *reheated*	0.3	0.1	0	Tr	Tr
279	**Tagliatelle with ham, mushroom and cheese**, chilled/frozen/longlife, *reheated*	0.3	0.1	Tr	0	Tr

[a] Contains 0.1g 22:5 per 100g food

153

VITAMIN D FRACTIONS

Analytical values for vitamin D (cholecalciferol) and 25-hydroxy vitamin D are given below, with total vitamin D activity taken as the sum of the cholecalciferol and five times the amount of 25-hydroxycholecalciferol. The amounts of vitamin D in the main tables have been interpolated from these values as well as those given in *Meat, Poultry and Game* (Chan *et al*, 1995).

Vitamin D fractions, μg per 100g food

No. 19-	Food	Vitamin D fractions		Total Vitamin D
		Vitamin D3	25-hydroxy Vitamin D$_3$	
	Bacon and ham			
1	**Bacon rashers, back**, *raw*	0.2	0.02	0.3
3	-, *grilled*	0.3	0.05	0.6
13	**middle**, *raw*	0.2	0.05	0.5
16	**streaky**, *raw*	0.5	0.08	0.9
18	-, *grilled*	0.4	0.06	0.7
	Burgers and grillsteaks			
28	**Beefburgers**, chilled/frozen, *raw*	0.2	0.19	1.2
	Meat pies and pastries			
69	**Steak and kidney/Beef pie**, individual, chilled/frozen, *baked*	0.3	0.07	0.7
	Sausages			
78	**Pork sausages**, chilled, *raw*	0.6	0.06	0.9
	Other meat products			
113	**Black pudding**, *raw*	0.6	Tr	0.6
116	**Chicken in crumbs**, stuffed with cheese and vegetables, chilled/frozen, *baked*	0.1	0.08	0.5
128	**Corned beef**, canned	0.3	0.20	1.3
131	**Faggots in gravy**, chilled/ frozen, *reheated*	Tr	0.09	0.5

No. 19-	Food	Vitamin D fractions		Total Vitamin D
		Vitamin D3	25-hydroxy Vitamin D_3	
Meat dishes				
143	**Pâté, liver**	0.8	0.07	1.2
248	**Moussaka**, chilled/frozen/ longlife, *reheated*	Tr	0.06	0.3
279	**Tagliatelle with ham,** mushroom and cheese, chilled/frozen/longlife, *reheated*	0.1	0.04	0.3

REFERENCES TO TABLES

1. Cashel, K., English, R. and Lewis, J. (1989) *Composition of Foods, Australia. Volume 1.* Department of Community Services and Health, Canberra

2. Voorlichtingsbureau voor de Voeding (1993) *Nevo Tabel. Nederlands voedingsstoffenbestand.* Den Haag

3. Feinberg, M., Favier, J.C. and Ireland-Ripert, J (1991) *Répertoire Général des Aliments. Table de Composition.* Institut National de la Recherche Agronomique, Technique et Documentation, Lavoisier

4. Souci Fachmann Kraut (1990) *Food composition and nutrition tables 1989/90, 4th edition.* Wissenschaftliche Verlagsgesellschaft mbH, Stuttgart

5. Carnovale E. and Miuccio, F.C. (1980) *Tabelle di Composizione Degli Alimenti.* Ministero dell'Agricoltura e delle Foreste. Istituto Nazionale della Nutrizione, Rome

6. Richardson, M., Posati, L.P., and Anderson, B.A. (1980) *Composition of foods: sausages and luncheon meats, raw, processed and prepared.* Agriculture Handbook No. 8-7, US Department of Agriculture, Washington D.C.

7. Anderson, B.A., Lauderdale, J.L., Hoke, I.M. (1986) *Composition of foods: beef products, raw, processed and prepared.* Agriculture Handbook No. 8-13, US Department of Agriculture, Washington D.C.

FOOD INDEX

Foods are indexed by their food number and **not** by their page number.

This index includes three kinds of cross-reference. The *first* is the normal coverage of alternative names (e.g. Back bacon see **Bacon rashers, back**). The *second* is to common examples of components of generically described foods, including brand names, which although not part of the food name have in general been included in the product description (e.g. Cumberland sausage see **Sausages, premium**). The *third* is for foods related to those in this book but whose nutritional value has already been covered in other supplements and are not repeated here (e.g. Baked beans with pork sausages, canned in tomato sauce see *Vegetables, Herbs and Spices* suppl.). Full references to these supplements are given on pages 6 and 7.

Back bacon   see **Bacon rashers, back**
Bacon   1-19
Bacon and liver, fried   see **Liver and bacon, fried**
Bacon loin steaks, grilled   19
Bacon rashers, back, dry-cured, grilled   6
   back, dry-fried   2
   back, fat trimmed, grilled   8
   back, fat trimmed, raw   7
   back, grilled   3
   back, grilled crispy   4
   back, microwaved   5
   back, raw   1
   back, reduced salt, grilled   9
   back, smoked, grilled   10
   back, sweetcure, grilled   11
   back, 'tendersweet', grilled   12
   middle, fried   14
   middle, grilled   15
   middle, raw   13
   streaky, fried   17
   streaky, grilled   18
   streaky, raw   16
Baked beans with burgers, canned in tomato sauce   see *Vegetables, Herbs and Spices* suppl.
Baked beans with pork sausages, canned in tomato sauce   see *Vegetables, Herbs and Spices* suppl.
Barbecued chicken wings   see **Chicken wings, marinated, barbecued**
Barbecued pork spare ribs   see **Pork spare ribs, 'barbecue style'**

Beef and pork sausages   see **Sausages, pork and beef**
Beef and spinach curry   160
Beefburgers, chilled/frozen, fried   29
Beefburgers, chilled/frozen, grilled   30
Beefburgers, chilled/frozen, raw   28
Beefburgers, economy   see **Economy burgers**
Beefburgers, homemade, fried   31
Beefburgers, homemade, fried, with bun   32
Beefburgers, homemade, grilled   33
Beefburgers, homemade, grilled, with bun   34
Beefburgers in gravy, canned   38
Beefburgers, low fat, chilled/frozen, fried   36
Beefburgers, low fat, chilled/frozen, grilled   37
Beefburgers, low fat, chilled/frozen, raw   35
Beef bourguignonne, chilled/frozen   see **Beef in sauce with vegetables**
Beef bourguignonne, homemade   161
Beef bourguignonne, homemade, made with lean beef   162
Beef carbonnade   see **Carbonnade of beef**
Beef casserole, canned   163
Beef casserole, made with canned cook-in sauce 164
Beef chow mein, retail, reheated   165
Beef curry, canned   168
Beef curry, chilled/frozen, reheated   169
Beef curry, chilled/frozen, reheated, with rice   170
Beef curry, homemade   166
Beef curry, reduced fat, homemade   167
Beef enchiladas   171
Beef goulash   see **Goulash**

Beef in sauce with vegetables, chilled/frozen, reheated 172

Beef/lamb hot pot   see **Lamb/Beef hot pot**

Beef kheema 173

Beef olives 174

Beef pancakes   see **Pancakes, beef**

Beef pie   see also **Steak and kidney pie**

Beef pie, chilled/frozen, baked 51

Beef pie, individual, chilled/frozen, baked 69

Beef sausages   see **Sausages, beef**

Beef slices 112

Beef spread   see **Meat spread**

Beef steak pudding, homemade 52

Beef stew 175

Beef stew and dumplings 177

Beef stew and dumplings, canned 178

Beef stew and dumplings, retail, cooked 179

Beef stew, made with lean beef 176

Beef, stir-fried with green peppers 180

Beef Stroganoff 181

Beef Wellington 182

Bierwurst 96

Big Mac 39

Biryani, lamb   see **Lamb biryani**

Black pudding, dry-fried 114

Black pudding, raw 113

Boeuf Stroganoff   see **Beef Stroganoff**

Bolognese sauce 183

    see also **Spaghetti bolognese**

Bratwurst 97

Brawn 115

Bridie/Scotch pie, individual 53

Burgers and grillsteaks 28-50

Burgers, beef   see **Beefburgers**

  chicken   see **Chicken burger, takeaway**

  economy   see **Economy burgers**

  hamburger, takeaway 47

    see also **Big Mac**, **Cheeseburger**, **Quarterpounder** and **Whopper burger**

Cabbage leaves, stuffed   see **Stuffed cabbage leaves**

Cannelloni, chilled/frozen, reheated 184

Carbonnade of beef 185

Casserole, beef   see **Beef casserole** or **Beef stew**

  lamb's heart   see **Lamb's heart casserole**

  pork   see **Pork casserole, made with canned cook-in sauce**

  pork and apple   see **Pork and apple casserole**

  rabbit   see **Rabbit casserole**

Casserole, sausage   see **Sausage casserole**

Casseroled chicken   see *Meat, Poultry and Game suppl.*

Cervelat 98

Cheeseburger, takeaway 40

Chicken and ham pie   see **Chicken pie**

Chicken and mushroom pie, single crust, homemade 54

    see also **Chicken pie**

Chicken and pork chow mein   see **Pork and chicken chow mein**

Chicken breast in crumbs, chilled, fried 118

Chicken breast in crumbs, chilled, grilled 119

Chicken breast in crumbs, chilled, raw 117

Chicken breast, marinated with garlic and herbs, chilled/frozen, baked 120

Chicken burger, takeway 41

Chicken chasseur 186

Chicken chasseur, weighed with bone 187

Chicken curry, chilled/frozen, reheated 188

Chicken curry, chilled/frozen, reheated, with rice 189

Chicken curry, made with canned curry sauce 190

Chicken fingers, baked 121

Chicken fricassée 191

Chicken fricassée, reduced fat 192

Chicken in crumbs, stuffed with cheese and vegetables, chilled/frozen, baked 116

Chicken goujons, chilled/frozen, baked 122

Chicken in sauce with vegetables, chilled/frozen, reheated 193

Chicken in white sauce, canned 194

Chicken in white sauce, made with semi-skimmed milk 196

Chicken in white sauce, made with whole milk 195

Chicken kiev, frozen, baked 123

Chicken korma 197

    see also **Chicken curry**

Chicken nuggets, takeaway 124

Chicken pancakes   see **Pancakes, chicken**

Chicken pie, individual, chilled/frozen, baked 55

Chicken risotto 198

Chicken roll 125

Chicken slices 126

Chicken, stir-fried with mushrooms and cashew nuts 199

Chicken, stir-fried with peppers in black bean sauce 200

Chicken, stir-fried with rice and vegetables, frozen, reheated 201

Chicken tandoori, chilled, reheated 127

Chicken tikka masala   see **Chicken curry**

Chicken vindaloo 202
Chicken vindaloo, reduced fat 203
Chicken wings, marinated, chilled/frozen,
  barbecued 204
Chicken wings, marinated, chilled/frozen,
  barbecued, weighed with bone 205
Chilli con carne, homemade 206
Chilli con carne, canned 207
Chilli con carne, chilled/frozen, reheated 208
Chilli con carne, chilled/frozen, reheated, with
  rice 209
Chinese luncheon meat   see **Luncheon meat,
                              Chinese**
Chops, pork, in mustard and cream see **Pork chops
                              in mustard and cream**
Chorizo 99
Chow mein, beef   see **Beef chow mein**
Chow mein, pork and chicken   see **Pork and
                              chicken chow mein**
Continental style sausages 96-111
Coq au vin 210
Coq au vin, weighed with bone 211
Corned beef hash 212
Corned beef, canned 128
Cornish pastie, homemade 57
Cornish pastie, retail 56
Coronation chicken 213
Coronation chicken, reduced fat 214
Cottage pie 215
Cottage/Shepherd's pie, chilled/frozen,
  reheated 216
Cumberland sausage   see **Sausage, premium**
Cured pork shoulder   see **Pork shoulder, cured,
                              slices**
Curry, beef 166-170
                    see also **Beef kheema**
  beef and spinach   see **Beef and spinach curry**
  chicken   see **Chicken curry**
    see also **Chicken korma** and **Chicken vindaloo**
  lamb   see **Lamb curry**
       see also **Lamb kheema, Lamb rogan josh**
                        and **Lamb vindaloo**

Devilled kidneys 217
Doner kebab in pitta bread with salad 130
Doner kebab, meat only 129
Duck à l'orange, excluding fat and skin 218
Duck à l'orange, including fat and skin 219
Duck with pineapple 220

Economy burgers, frozen, grilled 43

Economy burgers, frozen, raw 42
Economy sausages   see **Sausages, economy**
Enchiladas, beef   see **Beef enchiladas**

Faggots in gravy, chilled/frozen, reheated 131
Frankfurter 100
Frankfurter with bun 101
Frankfurter with bun, ketchup, fried onions and
  mustard 102
Fricassée of chicken   see **Chicken fricassée**

Game pie 58
Gammon joint   see **Ham, gammon joint**
Gammon rashers   see **Ham, gammon rashers**
Garlic sausage 103
Goujons, chicken   see **Chicken goujons**
Goulash 221
        see also **Beef in sauce with vegetables**
Grillsteaks   see also **Beefburgers** and **Steaklets**
Grillsteaks, beef, chilled/frozen, fried 45
Grillsteaks, beef, chilled/frozen, grilled 46
Grillsteaks, beef, raw 44

Haggis, boiled 132
Ham 20-27
Ham 23
Ham and mushroom sauce with pasta   see **Pasta,
                        with ham and mushroom sauce**
Ham and pork, chopped, canned 133
Ham, canned 24
Ham, gammon joint, boiled 21
Ham, gammon joint, raw 20
Ham, gammon rashers, grilled 22
Ham, mushroom and cheese with tagliatelle   see
          Tagliatelle with ham, mushroom and cheese
Ham, Parma 25
Ham, premium 26
Ham shoulder slices   see **Pork shoulder slices**
Ham spread   see **Meat spread**
Hamburger, takeaway 47
Hamburgers   see also **Beefburgers**
Hash, corned beef   see **Corned beef hash**
Haslet, pork   see **Pork haslet**
Heart, lamb's, casseroled   see **Lamb's heart
                                      casserole**
Hot pot, lamb/beef   see **Lamb/beef hot pot with
                                      potatoes**
  Lancashire   see **Lancashire hot pot**
Hotdogs   see **Frankfurters**

Irish stew, homemade 222
Irish stew, canned 224
Irish stew, made with lean beef, homemade 223

Kabana 104
Kebab, doner see **Doner kebab**
  pork and pineapple see **Pork and pineapple kebabs**
  shish see **Shish kebab**
Kheema, beef see **Beef kheema**
  lamb see **Lamb kheema**
Kiev, chicken see **Chicken kiev**
Knackwurst 105
Koftas, lamb see **Lamb koftas**
Korma, chicken see **Chicken korma**

Lamb/Beef hot pot with potatoes, chilled/frozen,
  retail, reheated 231
Lamb biryani 225
Lamb biryani, reduced fat 226
Lamb curry, made with canned curry sauce 227
Lamb kheema 228
Lamb kheema, reduced fat 229
Lamb koftas 230
Lamb, minced, stewed see **Minced lamb, stewed**
Lamb roast, frozen, cooked 134
Lamb rogan josh 232
Lamb samosa, retail 59
Lamb samosa, homemade, baked 60
Lamb samosa, homemade, deep-fried 61
Lamb's heart casserole 235
Lamb, stir-fried with vegetables 233
Lamb vindaloo 234
Lancashire hot pot 236
  see also **Lamb/Beef hot pot**
Lasagne, homemade 237
Lasagne, chilled/frozen, reheated 238
Lemon chicken 239
Lincolnshire sausage see **Sausages, premium**
Liver and bacon, fried 240
Liver and onions, stewed 241
Liver pâté see **Pâté, liver**
Liver sausage 106
Liverwurst see **Liver sausage**
Loin steaks, bacon see **Bacon loin steaks**
Low fat beefburgers see **Beefburgers, low fat**
Low fat sausages see **Sausages, pork, reduced fat**
Luncheon meat, Chinese, steamed 136
Luncheon meat, canned 135

Meat and tomato sauce, with pasta see **Pasta with meat and tomato sauce**
Meat loaf, chilled/frozen, reheated 138
Meat loaf, homemade 137
Meat pâté, low fat see **Pâté, meat, low fat**
Meat pies and pastries 51-74
Meat spread 139
Meatballs see **Pork and beef meatballs in tomato sauce**
  see also **Lamb koftas**
Middle bacon see **Bacon rashers, middle**
Minced beef in gravy, canned 140
Minced beef, extra lean, stewed 243
Minced beef pie filling, canned 141
Minced beef, stewed 242
Minced beef with gravy with/without onions,
  canned 244
Minced beef with vegetables, stewed 245
Minced lamb, stewed 246
Mortadella 107
Moussaka 247
Moussaka, chilled/frozen/longlife, reheated 248

Nuggets, chicken see **Chicken nuggets, takeaway**

Pancakes, beef, frozen, shallow-fried 249
Pancakes, chicken, frozen, shallow-fried 250
Parma ham see **Ham, Parma**
Pasta bake, turkey see **Turkey pasta bake**
Pasta, with ham and mushroom sauce 251
Pasta, with meat and tomato sauce 252
Pastie, Cornish see **Cornish pastie**
Pastrami 142
Pâté, liver 143
Pâté, liver, in a tube 144
Pâté, meat, low fat 145
Pepperami 108
Peppers, stuffed see **Stuffed peppers**
Pie, beef see **Beef pie**
  chicken see **Chicken pie, individual**
  chicken and mushroom see **Chicken and mushroom pie, homemade**
  cottage see **Cottage pie**
  game see **Game pie**
  pork see **Pork pie**
  pork and egg ee **Pork and egg pie**
  shepherd's see **Shepherd's pie**
  steak and kidney see **Steak and kidney pie**
  turkey see **Turkey pie**

Pie filling, minced beef   see **Minced beef pie filling, canned**

Pheasant salmis   see **Salmis of pheasant**
Polony   109
Pork and apple casserole   253
Pork and beef meatballs in tomato sauce   254
Pork and beef sausages   see **Sausages, pork and beef**
Pork and chicken chow mein   255
Pork and egg pie   62
Pork and ham, chopped, canned   see **Ham and pork, chopped, canned**
Pork and pineapple kebabs   259
Pork casserole, made with canned cook-in sauce   256
Pork chops in mustard and cream   257
Pork chops in mustard and cream, weighed with bone   258
Pork haslet   146
Pork pie, individual   63
Pork pie, mini   64
Pork pie, sliced   65
Pork roast, frozen, cooked   147
Pork sausages   see **Sausages, pork**
Pork shoulder, cured, slices   27
Pork slices   148
Pork spare ribs, 'barbecue style', chilled/frozen, reheated   262
Pork spare ribs, 'barbecue style', chilled/frozen, reheated, weighed with bone   263
Pork spare ribs, 'barbecue style', homemade   260
Pork spare ribs, 'barbecue style', homemade, weighed with bone   261
Pork spare ribs in black bean sauce   264
Pork spare ribs in black bean sauce, weighed with bone   265
Pork, stir-fried with vegetables   266
Pork, sweet and sour   see **Sweet and sour pork**
Premium ham   see **Ham, premium**
Premium sausages   see **Sausages, premium**
Pudding, beef steak, homemade   see **Beef steak pudding, homemade**
   black   see **Black pudding**
   steak and kidney   see **Steak and kidney pudding**
   white   see **White pudding**

Quarterpounder, takeaway   48

Rabbit casserole   267
Ravioli, canned in tomato sauce   see *Cereals and Cereal Products* suppl.

Reduced fat beefburgers   see **Beefburgers, low fat**
Reduced fat sausages   see **Sausages, reduced fat**
Risotto, chicken   see **Chicken risotto**
Rissoles, savoury   149
Rogan josh, lamb   see **Lamb rogan josh**

Salami   110
Salmis of pheasant   268
Samosa, lamb   see **Lamb samosa**
Sauce, bolognese   see **Bolognese sauce**
Sausage casserole   269
Sausage rolls, flaky pastry, homemade   67
Sausage rolls, puff pastry   66
Sausage rolls, short pastry, homemade   68
Sausage, garlic   see **Garlic sausage**
Sausage, liver   see **Liver sausage**
Sausages   75-95
Sausages, beef, chilled, fried   76
Sausages, beef, chilled, grilled   77
Sausages, beef, chilled, raw   75
Sausages, pork and beef, chilled, grilled   88
Sausages, pork and beef, chilled, raw   87
Sausages, pork and beef, economy, chilled, fried   91
Sausages, pork and beef, economy, chilled, grilled   92
Sausages, pork and beef, economy, chilled, raw   90
Sausages, pork and beef, frozen, raw   89
Sausages, pork, chilled, fried   79
Sausages, pork, chilled, grilled   80
Sausages, pork, chilled, raw   78
Sausages, pork, frozen, fried   82
Sausages, pork, frozen, grilled   83
Sausages, pork, frozen, raw   81
Sausages, pork, reduced fat, chilled/frozen, fried   85
Sausages, pork, reduced fat, chilled/frozen, grilled   86
Sausages, pork, reduced fat, chilled/frozen, raw   84
Sausages, premium, chilled, fried   94
Sausages, premium, chilled, grilled   95
Sausages, premium, chilled, raw   93
Saveloy, unbattered, takeaway   111
Scotch pie   see **Bridie/Scotch pie, individual**
Schnitzel, Wiener   see **Wiener schnitzel**
Shepherd's pie   270
Shepherd's pie, chilled/frozen, reheated   216
Shish kebab in pitta bread with salad   151
Shish kebab, meat only   150
Shish kebab, with onions and peppers   271
Spaghetti bolognese   272
Spaghetti bolognese, canned   see *Cereals and Cereal Products* suppl.
Spaghetti bolognese, chilled/frozen, reheated   273

Spare ribs, pork see **Pork spare ribs**

Steak and kidney pie, double crust, homemade 71

Steak and kidney pie, single crust, homemade 70

Steak and kidney pudding, canned 72

Steak and kidney pudding, homemade 73

Steak and kidney/Beef pie, individual, chilled/frozen, baked 69

Steak pudding, homemade see **Beef steak pudding, homemade**

Steak, stewed with gravy see **Stewed steak with gravy**

Steaklets, frozen, raw 49

Stew, beef see **Beef stew**

Stew, Irish see **Irish stew**

Stewed steak with gravy, canned 152

Stir-fried, beef see **Beef, stir-fried with green peppers**

    chicken 199-201

    duck with pineapple see **Duck with pineapple**

    lamb see **Lamb stir-fried with vegetables**

    pork see **Pork, stir-fried with vegetables**

    turkey see **Turkey, stir-fried with vegetables**

Streaky bacon see **Bacon rashers, streaky**

Stroganoff, beef see **Beef Stroganoff**

Stuffed cabbage leaves 274

Stuffed peppers 275

Sweet and sour pork, homemade 276

Sweet and sour pork, made with canned sweet and sour sauce 278

Sweet and sour pork, made with lean pork, homemade 277

Sweetcure bacon see **Bacon rashers, back, sweetcure**

Tagliatelle with ham, mushroom and cheese, chilled/frozen/longlife, reheated 279

Takeaway, cheeseburger see **Cheeseburger**

    chicken burger see **Chicken burger**

    hamburger see **Hamburger**

    see also **Beefburger, homemade**

    Quarterpounder see **Quarterpounder**

Tandoori chicken see **Chicken tandoori, chilled, reheated**

Tendersweet bacon see **Bacon rashers, back, 'tendersweet'**

Toad in the hole 280

Toad in the hole, made with skimmed milk and reduced fat sausages 281

Tongue see also **Meat, Poultry and Game** suppl.

Tongue, canned 153

Tongue slices 154

Tripe and onions, stewed 282

Turkey and pasta bake 283

Turkey pie, single crust, homemade 74

Turkey roast, frozen, cooked 155

Turkey roll 156

Turkey slices 157

Turkey steaks in crumbs, frozen, grilled 158

Turkey, stir-fried with vegetables 284

Venison in red wine and port 285

Vindaloo, chicken see **Chicken vindaloo**

Vindaloo, lamb see **Lamb vindaloo**

White pudding 159

Whopper burger 50

Wiener schnitzel 286